CARDIAC ARRHYTHMIAS:
New Therapeutic Drugs
and Devices

DEVELOPMENTS IN CARDIOVASCULAR MEDICINE

Lancee, C.T., ed.: Echocardiology, 1979. ISBN 90-247-2209-8.
Baan, J., Arntzenius, A.C., Yellin, E.L., eds.: Cardiac dynamics. 1980. ISBN 90-247-2212-8.
Thalen, H.J.T., Meere, C.C., eds.: Fundamentals of cardiac pacing. 1970. ISBN 90-247-2245-4.
Kulbertus, H.E., Wellens, H.J.J., eds.: Sudden death. 1980. ISBN 90-247-2290-X.
Dreifus, L.S., Brest, A.N., eds.: Clinical applications of cardiovascular drugs. 1980. ISBN 90-247-2295-0.
Spencer, M.P., Reid, J.M., eds.: Cerebrovascular evaluation with Doppler ultrasound. 1981. ISBN 90-247-2348-1.
Zipes, D.P., Bailey, J.C. Elharrar, V., eds.: The slow inward current and cardiac arrhythmias. 1980. ISBN 90-247-2380-9.
Kesteloot, H., Joossens, J.V., eds.: Epidemiology of arterial blood pressure. 1980. ISBN 90-247-2386-8.
Wackers, F.J.T., ed.: Thallium-201 and technetium-99m-pyrophosphate myocardial imaging in the coronary care unit. 1980. ISBN 90-247-2396-5.
Maseri, A., Marchesi, C., Chierchia, S., Trivella, M.G., eds.: Coronary care units. 1981. ISBN 90-247-2456-2.
Morganroth, J. Moore, E.N., Dreifus, L.S., Michelson, E.L., eds.: The evaluation of new antiarrhythmic drugs. 1981. ISBN 90-247-2474-0.
Alboni, P.: Intraventricular conduction disturbances. 1981. ISBN 90-247-2483-X.
Rijsterborgh, H., ed.: Echocardiology. 1981. ISBN 90-247-2491-0.
Wagner, G.S., ed.: Myocardial infarction. Measurement and intervention. 1982. ISBN 90-247-2513-5.
Meltzer, R.S., Roelandt, J., eds.: Contrast echocardiography. 1982. ISBN 90-247-2531-3.
Amery, A., Fagard, R., Lijnen, R., Staessen, J., eds.: Hypertensive cardiovascular disease; pathophysiology and treatment. 1982. ISBN 90-247-2534-8.
Bouman, L.N., Jongsma, H.J., eds.: Cardiac rate and rhythm. 1982. ISBN 90-247-2626-3.
Morganroth, J., Moore, E.N., eds.: The evaluation of beta blocker and calcium antagonist drugs. 1982. ISBN 90-247-2642-5.
Rosenbaum, M.B., ed.: Frontiers of cardiac electrophysiology. 1982. ISBN 90-247-2663-8.
Roelandt, J., Hugenholtz, P.G., eds.: Long-term ambulatory electrocardiography. 1982. ISBN 90-247-2664-8.
Adgey, A.J., ed.: Acute phase of ischemic heart disease and myocardial infarction. 1982. ISBN 90-247-2675-1.
Hanrath, P., Bleifeld, W., Souquet, eds.: Cardiovascular diagnosis by ultrasound. Transesophageal, computerized, contrast, Doppler echocardiography. 1982. ISBN 90-247-2692-1.
Roelandt, J., ed.: The practice of M-mode and two-dimensional echocardiography. 1983. ISBN 90-247-2745-6.
Meyer, J., Schweizer, P., Erbel, R., eds.: Advances in noninvasive cardiology. 1983. ISBN 0-89838-576-8.
Morganroth, Joel, Moore, E.N., eds.: Sudden cardiac death and congestive heart failure: Diagnosis and treatment. 1983. ISBN 0-89838-580-6.
Perry, H.M., ed.: Lifelong management of hypertension. ISBN 0-89838-582-2.
Jaffe, E.A., ed.: Biology of endothelial cells. ISBN 0-89838-587-3.
Surawicz, B., Reddy, C.P., Prystowsky, E.N., eds.: Tachycardias. 1984. ISBN 0-89838-588-1.
Spencer, M.P., ed.: Cardiac Doppler diagnosis. ISBN 0-89838-591-1.
Villareal, H.V., Sambhi, M.P., eds.: Topics in pathophysiology of hypertension. ISBN 0-89838-595-4.
Messerli, F.H., ed.: Cardiovascular disease in the elderly. 1984. ISBN 0-89838-596-2.
Simoons, M.L., Reiber, J.H.C., eds.: Nuclear imaging in clinical cardiology. ISBN 0-89838-599-7.
Ter Keurs, H.E.D.J., Schipperheym, J.J., eds.: Cardiac left ventricular hypertrophy. ISBN 0-89838-612-8.
Sperelakis, N., ed.: Physiology and pathophysiology of the heart. ISBN 0-89838-615-2.
Messerli, F.H., ed.: Kidney in essential hypertension. 1983. ISBN 0-89838-616-0.
Sambhi, M.P., ed.: Fundamental fault in hypertension. ISBN 0-89838-638-1.
Marchesi, D., ed.: Ambulatory monitoring: Cardiovascular system and allied applications. ISBN 0-89838-642-X.
Kupper, W., Macalpin, R.N., Bleifeld, W., eds.: Coronary tone in ischemic heart disease. ISBN 0-89838-646-2.
Sperelakis, N., Caulfield, J.B., eds.: Calcium antagonists: Mechanisms of action on cardiac muscle and vascular smooth muscle. ISBN 0-89838-655-1.
Godfraind, T., Herman, A.S., Wellens, D., eds.: Entry blockers in cardiovascular and cerebral dysfunctions. ISBN 0-89838-658-6.
Morganroth, J., Moore, E.N., eds.: Interventions in the acute phase of myocardial infarction. ISBN 0-89838-659-4.
Abel, F.L., Newman, W.H., eds.: Functional aspects of the normal, hypertrophied, and failing heart. ISBN 0-89838-665-9.
Sideman, S., and Beyar, R., eds.: Simulation and imaging of the cardiac system. ISBN 0-89838-687-X.
van de Wall, E., Lie, K.I., eds.: Recent views on hypertrophic cardiomyopathy. ISBN 0-89838-694-2.
Beamish, R.E., Singal, P.K., Dhalla, N.S., eds.: Stress and heart disease. ISBN 089838-709-4.
Beamish, R.E., Panagia, V., Dhalla, N.S., eds.: Pathogenesis of stress-induced heart disease. ISBN 0-89838-710-8.

CARDIAC ARRHYTHMIAS:
New Therapeutic Drugs and Devices

Proceedings of the Symposium on
New Drugs and Devices, held at Philadelphia, PA
October 4 and 5, 1984

edited by

Joel Morganroth
Likoff Cardiovascular Institute
of Hahnemann Medical College and Hospital

and

E. Neil Moore
School of Veterinary Medicine
University of Pennsylvania

Martinus Nijhoff Publishing
a member of the Kluwer Academic Publishers Group
Boston/Dordrecht/Lancaster

Distributors for North America:
Kluwer Academic Publishers
190 Old Derby Street
Hingham, MA 02043

Distributors Outside North America:
Kluwer Academic Publishers Group
Distribution Centre
P.O. Box 322
3300 AH Dordrecht
THE NETHERLANDS

Library of Congress Cataloging in Publication Data

Symposium on New Drugs and Devices (5th : 1984:
Philadelphia, Pa.)
Cardiac arrhythmias.

(Developments in cardiovascular medicine)
1. Arrhythmia—Chemotherapy—Evaluation—Congresses.
2. Myocardial depressants—Congresses. 3. Arrhythmia—
Treatment—Instruments—Congresses. 4. Cardiovascular
instruments, Implanted—Congresses. I. Morganroth, Joel.
II. Moore, E. Neil. III. Title. IV. Series.
[DNLM: 1. Anti-Arrhythmia Agents—standards—congresses.
2. Anti-Arrhythmia Agents—therapeutic use—congresses.
3. Arrhythmia—therapy—congresses. 4. Pacemaker
Artificial—therapeutic use—congresses.
W1 DE997VME / QV 150 S98972 1984c]
RC685.A65S965 1984 616.1'28061 85-4846
ISBN 0-89838-716-7

Printed in the United States of America

CONTENTS

1984 Fifth Annual Symposium on New Drugs and Devices

FACULTY

J. Thomas Bigger, Jr., M.D.
Professor of Medicine and Pharmacology
Columbia University

Ronald W. F. Campbell, M.D.
Senior Lecturer in Cardiology and
Honorary Consultant Cardiologist
Freeman Hospital, England

Jay N. Cohn, M.D.
Professor of Medicine
Head, Cardiovascular Division
University of Minnesota Medical School

Stewart J. Ehrreich, Ph.D.
Deputy Director, Cardio-Renal Division
Food and Drug Administration

John D. Fisher, M.D.
Associate Professor of Medicine
Albert Einstein College of Medicine

Allan M. Greenspan, M.D.
Associate Professor of Medicine
Hahnemann University

Donald C. Harrison, M.D.
William G. Irwin Professor of Cardiology
Stanford University

Brian F. Hoffman, M.D.
David Hosack Professor
College of Physicians & Surgeons, Columbia University

Leonard N. Horowitz, M.D.
Associate Professor of Medicine
Hahnemann University

Raymond J. Lipicky, M.D.
Acting Director, Division of Cardio-Renal Drug Products
Food and Drug Administration

E. Neil Moore, D.V.M., Ph.D.
Associate Professor of Medicine
University of Pennsylvania

Joel Morganroth, M.D.
Professor of Medicine and Pharmacology
Hahnemann University

Gordon W. Pledger, Ph.D.
Group Leader, Statistical Evaluation Branch
Division of Biometrics
Food and Drug Administration

Craig M. Pratt, M.D.
Assistant Professor of Medicine
Baylor College of Medicine

Glenn A. Rahmoeller
Acting Director, Division of Cardiovascular Devices
Food and Drug Administration

John C. Somberg, M.D.
Assistant Professor of Medicine and Molecular Pharmacology
Albert Einstein College of Medicine

Scott R. Spielman, M.D.
Associate Professor of Medicine
Hahnemann University

Robert Temple, M.D.
Acting Director, Office of Drug Research & Review
Food and Drug Administration

August M. Watanabe, M.D.
Professor and Chairman, Department of Medicine
Professor of Pharmacology, Indiana University Medical Center

Roger A. Winkle, M.D.
Associate Professor of Medicine
Stanford University Medical Center

Raymond L. Woosley, M.D., Ph.D.
Associate Professor of Medicine and Pharmacology
Vanderbilt University

Douglas P. Zipes, M.D.
Professor of Medicine
Indiana University School of Medicine

PREFACE

In October, 1980 we convened our first Symposium on New Drugs to evaluate new antiarrhythmic agents. The participants comprised members of the Food and Drug administration, the academic community and the research divisions of the pharmaceutical industry. Five years ago, there were a host of new antiarrhythmic agents being evaluated in the United States in which basic research design questions and approaches were so unclear and confusing that we hoped the discussions generated would lend new guidlines to the evaluation process. We also believed that this Forum might help speed to market new antiarrhythmic agents to improve the drug armamentarium of physicians in the United States. Antiarrhythmic agents are used to treat patients with cardiac arrhythmias including those exhibiting symptoms from hemodynamic consequences due to life-threatening ventricular arrhythmias as well as patients with no symptoms from ventricular arrhythmias that increase the risk of sudden cardiac death. Treatment for this later group is offered with the hope that the abolition of the ventricular arrhythmias will be prophylactic against sudden death.

This fifth annual Symposium on New Drugs and Devices has returned again to its origin by selecting the topic of new antiarrhythmic agents and devices. Over the past five years no new antiarrhythmic agents have, in fact, been released although one appears to be imminent (tocainide) and three others have recently been recommended for approval by the Cardio-Renal Advisory Board (amiodarone, flecainide and mexiletine).

Many of the same questions raised five years ago were raised again at this meeting in 1984. Basic questions as to the role of animal models as a guide for selecting new antiarrhythmic drugs to the proper development of study designs for electrophysiologic, hemodynamic, and pharmacologic characteristics were again addressed. The status of specific antiarrhythmic agents as defined by a modification of the Vaughn Williams Classification with particular emphasis on research questions yet remaining was an important area in discussion. New ways of handling drug data, from remote data entry to new statistical approaches, with a comparison of the various detection models were also addressed. The differentiation of using arrhythmia suppression vs sudden cardiac death prevention as therapy endpoints was an important consideration.

For the first time we also discussed the evaluation of approaches to designing studies to investigate atrial arrhythmias and selecting proper efficacy endpoints as well as the role of the nervous system in their generation.

The emerging and exciting role of antiarrhythmic devices which include not only pacemakers to terminate ventricular tachyarrhythmias but those that take the form of internal cardioverters and defibrillators were valuable topics. A discussion as to new device regulation for such equipment and standards for determining efficacy and safety were important considerations.

The manuscripts in the proceedings were selected to provide an overview of a particular topic and the following group discussions represented viewpoints aired among all of the participants. While this symposium, as it did five years ago, could not provide a unanamious consensus on many of the issues, it clearly pointed to new important research questions and provided direction to both preclinical and clinical settings.

Joel Morganroth, M.D.
E. Neil Moore, D.V.M., Ph.D.

Philadelphia, PA, U.S.A.

I. BASIC CONSIDERATIONS

1

WHAT ANIMAL MODELS ARE USEFUL IN SELECTING NEW ANTIARRHYTHMIC DRUGS?

E. Neil Moore, D.V.M., Ph.D. and Joseph F. Spear, Ph.D.

Although many animal models are useful for evaluating new antiarrhythmic agents, it must be admitted that no animal model is ideal, nor can any single animal model alone predict antiarrhythmic efficacy in man. The selection of which animal model(s) to use depends first of all on whether one is "screening" for a new active compound or alternatively,"evaluating" an already identified active antiarrhythmic drug. This paper will deal primarily with models designed to evaluate mechanisms of action and effectiveness of new antiarrhythmic agents.

The animal models that we have found most useful in evaluating antiarrhythmic agents include, 1) cellular electrophysiological model, 2) acute coronary ligation models, 3) chronic myocardial infarct models, and 4) ventricular fibrillation threshold models. Other investigators have found drug arrhythmia models and neural arrhythmia models. In addition to being able to evaluate antiarrhythmic efficacy, many of these animal models can also provide useful information regarding side effects such as alterations in blood pressure, depression of cardiac contractility, CNS toxicity, and gastro-intestinal toxicity.

CELLULAR ELECTROPHYSIOLOGICAL MODELS

Most tachyarrhythmias in man are thought to be due to reentry. To prevent reentrant rhythms antiarrhythmic drugs must alter refractoriness, excitability and/or conduction velocity. By recording transmembrane potentials from normal canine Purkinje fibers one can evaluate effects of an agent on 1) refractoriness as predicted by action potential duration, 2) conduction velocity as correlated with rate of depolarization and 3) excitability as defined by the threshold current for excitation. Thus microelectrode studies in normal fibers can be important in predicting whether a new compound has the potential to prevent or terminate reentrant tachyarrhythmias.

In more recent years cellular electrophysiological studies have been extended to studying normal fibers made abnormal by the administration of drugs (digitalis, barium or cesium), or altered electrolyte (potassium, sodium and/or calcium) in the superfluent fluid bathing the preparation. These studies have demonstrated that arrhythmias can occur by mechanisms other than altering normal automaticity and/or reentry. Abnormal slow channel pacemaker activity has been demonstrated in depressed fibers as a result of calcium and sodium currents that become activated at low membrane potentials. These slow channel currents can be blocked by calcium channel blockers. In addition, early and late afterdepolarizations as well as triggered automaticity have been observed associated with digitalis toxicity and/or altered superfluent fluid bathing the preparation. The demonstration of triggered automaticity where programmed electrical stimulation can both start and stop a triggered automatic focus has truly complicated the analysis and treatment of arrhythmias in man. It had originally been thought that a reentrant arrhythmia in man could be identified by the ability to initiate and terminate the arrhythmia with programmed stimulation. Automatic rhythms were then considered to be ones that could neither be initiated nor terminated with programmed stimulation. Thus the recent demonstration of triggered automaticity in cellular electrophysiological models has further complicated the evaluation of cardiac dysrhythmias in man. Also, in infarcted cardiac tissues studied with microelectrodes both reentrant arrhythmias and triggered automatic foci have been identified in the same heart. This points out the difficulty of identifying the mechanism(s) for a dysrhythmia as well as the difficulty for any single agent to be effective when several different mechanisms for arrhythmogenesis may be present.

We believe that cellular electrophysiological models can be very helpful in evaluating new antiarrhythmic agents, particularly in evaluating antiarrhythmic agents that influence abnormal automatic activity. In fact, cellular electrophysiological techniques are the only ones that are able to directly demonstrate afterdepolarizations and triggered automaticity as well as to identify the ionic conductance changes responsible for antiarrhythmic activity.

VENTRICULAR FIBRILLATION THRESHOLD MODELS

Ventricular fibrillation is associated with electrical instability of the ventricles and is the primary cause of early sudden death after myocardial infarction. Although Wiggers and Wegria (2) discovered that an appropriately timed impulse of sufficient strength delivered to the ventricles in late systole could induce ventricular fibrillation, it remained for more sophisticated recording techniques to demonstrate the mechanism of how these currents induced fibrillation. Euler and Moore (3) showed by recording from the myocardium within 1 to 2 mm of the site of the current-delivering ventricular fibrillation threshold (VFT) electrode that the high VFT current resulted in fractionated continuous activity developing immediately surrounding the VFT current electrode. When recordings were made more than 1 cm away from the VFT current electrode, then continuous repetitive fractionated activity was not recorded. Thus the VFT method causes ventricular fibrillation by the production of microreentry in the myocardial tissue immediately surrounding the VFT electrode. The fractionated continuous activity observed with the VFT technique is grossly indistinguishable from that observed in infarcted hearts prone to the development of ventricular tachycardia and/or ventricular fibrillation (VT/VF).

The VFT technique as used in different laboratories has many variations including 1) delivery of a single pulse properly timed within the vulnerable period, 2) a train of constant current pulses delivered across the vulnerable period, 3) continuous 60 Hz current delivered until ventricular fibrillation develops and 4) transthoracic discharge. We believe that the latter two VFT methods have many difficulties and often provide variable results (1). The VFT short train of pulses positioned across the vulnerable period introduced by Han is the easiest and most reproducible technique of all the VFT methods (4) . With this technique it is necessary only to increase the current intensity of the train of pulses in a step wise fashion until fibrillation ensues. This procedure can be done quickly and allows one to study the effects of single variables on ventricular vulnerability in a controlled animal model system. Despite the fact that VFT is an artificial technique for inducing ventricular fibrillation via current injection, it nevertheless is a method that if appropriately controlled can provide clinically useful data as to the potential effectiveness of new antiarrhythmic agents.

ACUTE CANINE CORONARY LIGATION MODELS

It has been known since 1881 when Cohnheim (5) first occluded the circumflex branch of the left coronary artery in anesthetized dogs that coronary occlusion can result in ventricular fibrillation. In addition, in animals in which fibrillation does not develop during coronary occlusion, it often develops immediately upon release and reperfusion of the coronary artery. The problem with evaluating arrhythmic agents in acute occlusion-reperfusion coronary ligation models, is that there are many complicating factors that govern whether arrhythmias will or will not develop. Different laboratories have reported large variations in the percentage of animals which develop VF during occlusion and during release of the acute occlusion. In addition, it has recently been demonstrated that there are two periods of arrhythmias during acute occlusion, one at 2 to 12 minutes and another at 13 to 30 minutes after occlusion which further complicates the acute occlusion models (6). In our laboratory, we have seen variations between 35 to 85% of the animals which develop ventricular fibrillation upon occlusion and/or reperfusion. This variability is related to the fact that 1) the site of coronary ligation will vary, 2) the degree of manipulation of the coronary artery prior to ligation will vary with the investigator's expertise, 3) the duration of ligation will vary in different laboratories, 4) the heart rate and rhythm both preceding as well as during the acute ligation will vary, 5) the left ventricular function due to the amount of collateralization of the heart will vary 6) the neurohumoral state will vary with level of anesthesia and type of anesthetic used and 7) both the electrolyte and acid/base state will vary among animals. Also the effects of drugs given prior to versus after producing the coronary ligation will alter antiarrhythmic effectiveness as a result of the varying distribution of the drug within the myocardium. Further variability in coronary ligation models can result from species variations. For example in the pig there are practically no collateral coronary arteries. Thus in the pig nearly 100% of the animals will undergo VF following occlusion of the LAD. Although the acute occlusion-reperfusion models can provide useful information we would not select them as our first models for evaluating antiarrhythmic agents due to the wide variability in the frequency of arrhythmias that occur with occlusion-reperfusion. In addition, occlusion-reperfusion studies necessitate a large number of animals being utilized in order to obtain statistical significance for reliable antiarrhythmic efficacy.

SUBACUTE CORONARY LIGATION MODELS

Harris and Rojas (7) demonstrated over 30 years ago that when if the left anterior descending (LAD) coronary artery is ligated in a 2-stage procedure that most dogs survive the acute procedure and then 15 to 24 hours after the occlusion there is a high incidence of spontaneous ventricular arrhythmias. These arrhythmias last for up to 72 hours and are rarely accompanied by ventricular fibrillation. We believe that this model is quite a severe test of new antiarrhythmic agents primarily because drug levels necessary to suppress ectopic rhythms in this model have been higher than those found subsequently to be required for suppression of ectopic activity in man (8). This may relate to differences in mechanisms underlying the arrhythmias, to differences in metabolism of the compound as well as to other species differences. The Harris subacute LAD ligation model has multiple mechanisms for giving rise to arrhythmias including altered automaticity within the Purkinje system, triggered automaticity and reentry as initiated by programmed stimulation. One potential problem with the Harris model is that the percentage of beats that are ectopic and abnormal can vary from animal to animal. It is not known whether it is a more severe test of a drug to investigate it in an animal that has 100% of the beats which are ectopic versus an animal in which only 70% are abnormal. Variability in ectopic frequency requires statistical evaluation when studying any new antiarrhythmic agent. However, the Harris 2-stage coronary ligation model does have the advantage that studies can be done in conscious animals and that oral as well as I.V. doses of the new compound can be evaluated. Furthermore, potential side effects of the drug can be evaluated by recording blood pressure, cardiac contractility, and noting CNS and gastrointestinal toxicity. The extensive experience and background data on the effectiveness of commonly used antiarrhythmic drugs in this model makes it an excellent primary model in which to evaluate new antiarrhythmic agents.

CHRONIC ANIMAL MODELS

One problem that has faced experimental electrophysiologists for years has been the lack of an appropriate animal model in which malignant ventricular tachyarrhythmias could be induced by the same programmed electrical stimulation protocols as used in man having coronary heart disease. Several laboratories have recently discovered that the occlusion of the proximal left anterior descending coronary artery for a period of several hours followed by release and reperfusion will render canine hearts much more highly susceptible to the initiation of sustained ventricular tachyarrhythmias than hearts in which the coronary artery occlusion is maintained. The occlusion-reperfusion procedure results in a mottled heterogeneous infarct in which normal tissue is interspersed with dead tissue. It has been found that 3 to 5 days after the initial occlusion period that the same programmed electrical stimulation protocols as used in man will both initiate and terminate ventricular tachycardia and/or ventricular fibrillation (VT/VF). Programmed electrical stimulation studies must be done at least 3 days following the LAD surgical procedures in order to alleviate the complexities that are introduced by the spontaneous ventricular dysrhythmias that occur due to altered automaticity. A major advantage of this chronic myocardial infarct model is that the majority of animals recover from the LAD occlusion and that more than half of the infarct animals will be susceptible to electrically induced tachyarrhythmias. Another advantage of these chronic canine infarct models is that studies can be done in either unanesthetized or anesthetized animals and it is possible to give drugs orally as well as intravenously. Also the size, age, location and type of infarct is known and electrode recording and stimulating positions in relationship to the infarct can be characterized at post-mortem by routine staining methods. Since all infarcts are due to ischemia resulting from coronary occlusion-reperfusionexperimental population becomes more homogeneous than is often found in human studies where arrhythmias are associated not only with coronary artery disease but also with mitral valve prolapse, cardiomyopathies and congenital heart problems.

We have recently used computer-assisted signal averaging techniques to evaluate the presence of low level late potentials in dogs prone to ventricular tachyarrhythmias. We were pleased to see that not only can low level late potentials be observed in dogs more than 2 years after the LAD occlusion-reperfusion surgery but furthermore that only in dogs having late potentials was it possible to initiate ventricular tachyarrhythmias using programmed electrical stimulation. An example of a normal control dog signal averaged high gain ECG as contrasted to the recording of late potentials in a dog having VT initiated by programmed electrical stimulation is shown in Figure 1. Thus although these infarcts must undergo progressive healing over a number of weeks it is interesting to find that it still is possible to induce VT and VF in animals several years after their initial infarct. Two major differences between our chronic canine infarct animals and people having VT/VF is that infarction in the animals is not due to coronary artery disease and the incidence of spontaneous PVC's, VT and VF is much less frequent in the infarct animal.

We believe that these chronic myocardial infarct animals are probably the most useful model that we have for evaluating new antiarrhythmic agents. Most of the cardioactive agents that are commonly used in man have similar effects against the electrically evoked ventricular tachyarrhythmias in this animal model as observed in man. Also the dose levels for most compounds are very similar. It must be realized however, that the metabolism of new compounds may differ in the dog and man and thus effective metabolites may be missed. There also may be differences in the proarrhythmia effects of a new antiarrhythmic compound in man versus our chronic canine infarct model. However, the observation that we have seen lidocaine make arrhythmias worse in this dog model in a similar manner to that observed in man suggests that this model may be helpful in predicting proarrhythmia effects. Also, desirable or undesirable interactions between drugs can be evaluated. Besides having a more homogeneous population prone to the development of ventricular tachyarrhythmias, our animal models also permit more uniform experimental protocols to be used. For example in man, it is difficult and probably unethical to repeatedly use a programmed electrical stimulation sequence to validate the reproducibility of ventricular fibrillation in a potential sudden cardiac death victim. Thus the reproducibilty of a given programmed pacing sequence to reproducibly produce VF or hemodynamically significant VT is not possible in most human studies. In our chronic infarct animals, on the other hand, it is possible to initiate VF 2 or 3 times to demonstrate definite reproducibility of the induction of fibrillation with a given pacing protocol. This adds significance to programmed electrical stimulation (PES) results. Also, we can use programmed electrical pacing protocols in experimental studies using our chornic infarct animal model that are difficult to perform in man.

We have found that results with programmed electrical stimulation can vary depending upon differences in the parameters of in the electrical pacing protocol. For example, we have found that the 1) site of stimulation, 2) stimulus duration, 3) stimulus intensity, 4) stimulus polarity, 5) number of extrastimuli, 6) number of basic stimuli and 7) the basic stimulus rate can all influence whether a ventricular tachyarrhythmia is reproducible initiated. Development of chronic canine infarct models to test PES should help in developing more reproducible electrical stimulation protocols.

SUMMARY

In summary, there are many animal models that are useful in selecting new antiarrhythmic drugs. The selection of which model is most idea depends upon precisely what question is being asked. The large number of experimental models used to evaluate antiarrhythmic compounds points out the inability of any one model to define the probability of antiarrhythmic efficacy in man. It has therefore become standard practice to utilize a batter of animal models for the evaluation of new antiarrhythmic agents. Each model has its own advantages and disadvantages and it is necessary to understand each model fully in oder to evaluate experimental findings and apply them to clinical settings.

We believe that the availability of the chronic myocardial infarction ventricular tachyarrhythmia model provides 1) an excellent opportunity to more precisely understand arrhythmia mechanisms, 2) to develop new techniques such as signal averaging for evaluating late low level potentials identifying hearts at high risk of sudden death 3) to identify new antifibrillatory drugs versus drugs that are effective primarily against PVC's and ventricular tachycardia 4) to identify new surgical techniques to eliminate VT/VF, and 5) to evaluate new pacing modalities including implantable cardioverters. Although all animal models are wrong, many are very useful in furthering our knowledge directed at decreasing the distressingly high mortality from heart disease.

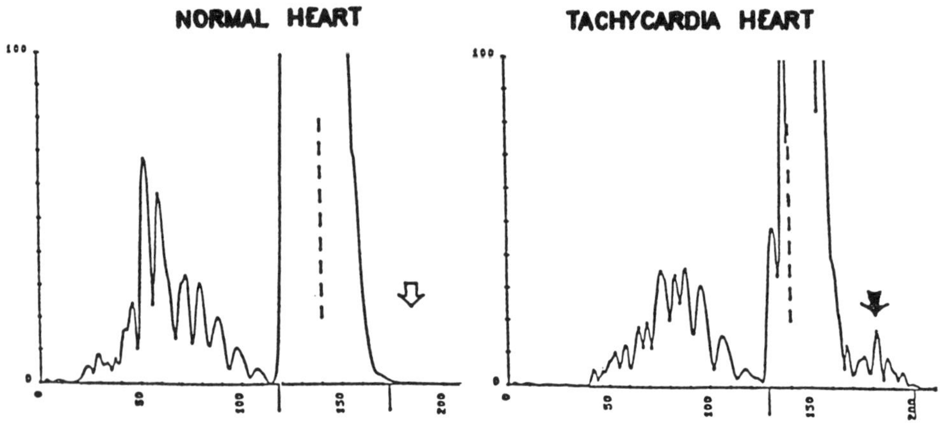

FIGURE 1: On the left hand is a computer processed, signal averaged electrocardiogram obtained in a normal animal. Note that the duration of the

QRS complex (dotted vertical line) is brief and that no potentials exist after the end of the QRS complex. The absence of any abnormal potentials is indicated by the open arrow. In contrast, in chronic canine infarct animal in which ventricular tachycardia could be reproducibly initiated with programmed electrical stimulation, it can be noted that there is a late low voltage potential (tail potential") which follws the end of the QRS complex. This late potential is emphasized by the solid arrow. We have never observed any late low voltage potentials in any heart without tachyarrhythmias. The amplitude of these abnormal late potentials is exceedingly small being in the 2 to 15 microvolt range. Normally such low voltage potentials would never be observed since standard ECG's are recorded at 1 millivolt sensitivity.

REFERENCES

1. Moore EN and Spear JF: Acute animal models for the study of antiarrhythmic drugs for the prevention of sudden coronary death. In: Clinical Pharmacology of Antiarrhythmic Therapy. Ed. Lucchesi, BR, Dingell, JV and Schwarz RP Jr. Raven Press, N.Y. pp. 31-46, 1984.

2. Wiggers, CJ, and Wegria R: Ventricular fibrillation due to a single localized induction and condenser shock applied during the vulnerable phase of ventricular systole. Am. J. Physiol. 128:500-505.

3. Euler DE and Moore EN: Continuous fractionated electrical activity after stimulation of the ventricles during the vulnerable period: evidence for local reentry. Am. J. Cardiol. 46:783-791, 1980.

4. Han J: Ventricular vulnerability during acute coronary occlusion. Am. J Cardiol. 24:857-864, 1969.

5. Greenberg, HM and Dwyer, EM. Eds. Sudden Coronary Death Ann. NY Acad. Sci. 382:1-484, 1982.

6. Kaplinsky, E, Ogawa S, Balke CW and Dreifus LS: Two periods of early ventricular arrhythmia in the canine acute myocardial infarction model. Circ. 60:397-403, 1979.

7. Harris, AS and Rojas, AG: Initiation of ventricular fibrillation due to coronary occlusion. Exp. Med. Surg. 1:105-122, 1943.

8. Moore, EN, Spear, JF, Feldman HS and Moller R: Electrophysiological properties of a new antiarrhythmic drug - Tocainide. Am. J. Cardiol. 41:703-708, 1978.

2

HOW TO TEST FOR ANTIARRHYTHMIC DRUGS

BRIAN F. HOFFMAN, M.D.

I have been asked to consider several related questions. First: When during drug development and testing should electrophysiologic data be acquired? Second, what methods should be used to acquire this data? And, finally, during screening should one employ models of arrhythmias caused by ischemia or models of other sorts?

Unfortunately, there are no categorical answers to any of these questions, since the questions themselves are too imprecise to permit rational evaluation. To consider first the question of when should one make measurements of drug effects on electrophysiologic properties of cardiac tissues, I suggest that studies on drug-induced changes in the transmembrane potentials or sarcolemmal ion currents of cardiac fibers rationally should be the first biological test after completion of chemical synthesis. This sequence would permit one to determine whether or not each new agent brought about changes in electrical activity that, judged in terms of prior experience, were or were not likely to exert antiarrhythmic efficacy. One could determine whether or not a new agent blocked fast channels and, if so, the rate constants for use-dependent enhancement of block and for recovery from use-dependent block.[1,2,3,4] One could identify blockade of slow inward channels and its frequency-dependence[5] and quantify effects on the repolarizing current i_x[6] and the Purkinje fiber pacemaker current i_f.[7] These actions on ion channels could be correlated with observed changes in action potential configuration, excitability, conduction velocity, refractoriness, and automaticity. By comparison with data for other agents that were known to be clinically effective, one could make predictions about the probable efficacy of the new compound.

If, among a series of similar compounds, significant differences were found in terms of effects on specific ion channels, the information might be of great value in helping determine structure activity relationships. For example, there is good reason to believe that a drug that combines very rapidly with inactivated fast channels but dissociates very slowly from such channels has a high potential for toxicity. There is good reason to believe that a drug that signficantly attenuates the repolarizing current, i_x, has a significant potential to produce early afterdepolarizations and the arrhythmia torsades de pointes whenever the heart rate slows sufficiently. A drug that blocks the pacemaker current in Purkinje fibers, i_f, will have predictable effects on the capability of the His-Purkinje systems to develop an automatic rhythm, and very likely will exert the same effect on some classes of subsidiary atrial pacemakers. A drug that significantly changes the kinetics or voltage-dependence of inactivation of i_x, most likely will influence the abnormal automaticity that occurs at reduced levels of transmembrane potential and very likely will strongly influence sinus node impulse generation.[8] These and many other predictions about probable effects on the *in situ* heart could be made from studies of the effects of new compounds on electrical activity of isolated preparations of cardiac tissue.

In addition, it is possible to study drug effects on the several postulated arrhythmogenic mechanisms through the use of isolated preparations of cardiac tissue and intracellular microelectrodes. The mechanisms for impulse generation - normal automaticity, abnormal automaticity, and early and delayed afterdepolarizations can be evaluated in a variety of well-tested models. Abnormal automaticity can be induced quite reproducibly in Purkinje fibers and atrial and ventricular muscle either by exposure to low concentrations of barium[9] or by depolarizing current. Delayed afterdepolarizations can be induced in Purkinje cells or muscle fibers by exposure to digitalis[10] and in fibers of the coronary sinus by exposure to catecholamines. Early afterdepolarizations can be induced at slow rates of stimulation either by cesium or quinidine.[11,12] Thus the ability of a new chemical entity to attenuate or block triggered rhythms due to delayed afterdepolarizations could be evaluated in the cellular electrophysiologic study either from measurements of transmembrane potentials or measurement of the transient inward current[13] during voltage clamp. Similarly, the ability of a drug to slow or terminate abnormal automaticity could be evaluated in Purkinje fibers exposed to low concentrations of barium salts or Purkinje fibers partially depolarized by current clamp. Effects of a new compound on early afterdepolarizations could be studied in fibers exposed to cesium, N-acetylprocainamide[14] or a variety of other agents.

In summary, measurements of drug effects on the electrical activity of normal cardiac cells, studied with standard microelectrode and voltage-clamp techniques, would be sufficient to identify and quantify any induced changes in many of the ionic currents that underlie normal and abnormal electrical activity. It probably would be advisable to make some of the tests on two different types of cardiac tissue as Purkinje fibers and muscle fibers may respond differently to anti-arrhythmic drugs. This information, coupled with information on changes in the voltage time-course of the transmembrane potential, on conduction, on refractoriness, and on specific arrhythmogenic mechanisms, if evaluated in terms of prior experience, should permit reasonable predictions about anti-arrhythmic efficacy.

The next logical step in terms of this approach would be to test the drugs, and thus the predictions, in suitable animal models. If the predictions were validated by the response of the animal models, the information would be of great value to the chemist. The first step, often left to last, is to measure the effects of the new drug on the electrical activity of the intact, normal canine heart. Changes in P-R interval and sinus rate will provide important information not available through the standard cellular electrophysiological studies. Changes in H-V interval, QRS duration and J-T interval (corrected for rate) will either corroborate the data from studies on isolated tissues or provide new findings. Measurements of these intervals at different rates will provide a check on estimates of use-dependent blocking actions obtained from the isolated preparations.

The weaknesss of the sequence of course, results from our very limited understanding of the mechanisms for arrhythmias in the animal models usually employed and also by the fact that in most of the animal models the occurrence of arrhythmia is associated with ill-defined or unappreciated abnormalities of cellular electrophysiology or electrophysiology and fine structure. In the Harris dog, a model probably used more widely than any other to study efficacy against ventricular arrhythmias, our own data strongly suports the conclusion that the primary arrhythmogenic mechanism at 24 hours is abnormal automaticity. We have come to this conclusion both from studies on transmembrane potentials of tissue removed from 24-hour infarcts and from studies on the electrophysiology of the arrhythmic <u>in situ</u> heart. Typically, the rhythm is reproducibly captured by pacing at a rate slightly greater than the rate of the arrhythmia and after the end of overdrive there is no significant suppression of arrhythmic activity. Others, from similar studies, have concluded that delayed afterdepolarizations are the important mechanism.[15] In fact, in many animals it is likely that, even though abnormal automaticity is the dominant mechanism, both reentry and triggering contribute to the polymorphic or chaotic rhythm. A different kind of variability can be seen if one studies infarcted hearts approximately one week after coronary artery occlusion. Here the crucial variable is the anatomical extent of the infarct. In some dogs, as was shown many years ago,[16] the infarct is almost transmural and the only surviving cells are a thin shell of subendocardial Purkinje fibers. In other dogs, the infarct does not extend to the epicardium and a shell of epicardial muscle survives. In still other dogs, the infarct is patchy, and strands of viable muscle may extend here and there between epicardium and endocardium. If one considers the first and second models, one can anticipate quite different effects of the same drug on electrophysiologic properties. In the case of the surviving subendocardial Purkinje fibers, action potential duration is greatly prolonged in the areas subtended by the infarct. The cells thus are suitable for initiation of reentrant rhythms. As has been shown, some antiarrhythmic drugs decrease the dispersion of refractoriness by prolonging action potential duration much more in normal cells than in the cells damaged by the ischemic event.[17] A sharply contrasting condition is provided by the heart in which a layer of subepicardial cells survives the ischemic insult. In this tissue, the transmembrane action potential shows normal values for resting potential, action potential amplitude and V_{max}, but a marked decrease in action potential duration. If the same drug were to be used and were to increase action potential duration more in the normal cells than in the cells damaged by ischemia, an increase in the dispersion of refractoriness would result. If one employs instead a model of induced or spontaneous arrhythmias occurring late after infarction, similar variability of arrhythmogenic mechanism is encountered.

Obviously the use of animal models of arrhythmias to test the predictions made from studies on cellular electrophysiology will not be useful unless, for each animal model, the arrhythmogenic mechanism is known and is consistent. The problem is made more complicated by the observation that a single drug-induced change in cardiac electrophysiologic properties may be sufficient to exert antiarrhythmic action against arrhythmias caused by a variety of mechanisms. As one example, it is instructive to ask what antiarrhythmic actions might result from partial block of fast inward channels. Obviously, if the blockade of fast channels decreased the safety factor for propagation of the fast response, a reentrant rhythm in a standard model might be terminated because the action potential in the reentrant circuit might lose its ability to stimulate distal to the area of slow conduction and one-way block. Partial blockade of fast inward channels also might terminate an arrhythmia caused by delayed afterdepolarizations. These are thought to occur when calcium overload results in an oscillatory change in intracellular calcium-ion activity and this, in turn, opens the transient inward current channel.[13] One of the factors that regulates intracellular calcium activity is intracellular sodium concentration.[18] Partial block of fast inward channels would cause a reduction in net sodium entry with each action potential, and this, through the sodium-calcium exchanger, might permit a reduction in intracellular calcium. This should, in turn, reduce the magnitude of the transient of the inward current and the amplitude of the delayed afterdepolarizations. If the after-depolarizations became sufficiently reduced in amplitude, they would fail to trigger action potentials and the arrhythmia would cease. Partial blockade of fast channels also could slow or terminate an arrhythmia due to abnormal automaticity if inward current in slowly inactivating channels[19] or window current[20] contributed appreciably to the diastolic depolarization. I mention these problems because one might argue that because the mechanisms are not well known for most animal models, one might as well go directly to tests against real arrhythmias in human subjects. This procedure clearly would be able to identify potentially useful antiarrhythmic drugs, but would do little or nothing to improve the science of antiarrhythmic drug design.

REFERENCES

1 Hondeghem LM, Katzung BG: Time- and voltage-dependent interactions of antiarrhythmic drugs with cardiac sodium channels. Biochim Biophys Acta 472:373-398, 1977.

2. Courtney KR: Interval-dependent effects of small antiarrhythmic drugs on excitability of guinea pig myocardium. J Mol Cell Cardiol 12:1273-1273, 1980.

3. Gintant GA, Hoffman BF: Use-dependent block of cardiac sodium channels by quaternary derivatives of lidocaine. Pflugers Arch 400:121-129, 1984.

4. Bean BP, Cohen CJ, Tsien RW: Lidocaine binding to resting and inactivated cardiac sodium channels. Biophys J 33: 208, 1981.

5. Trautwein W, Pelzer D, McDonald TF, Osterrieder W: AQA 39, a bradycardic agent which blocks myocardial slow inward channels in a frequency- and voltage-dependent manner. Naunyn-Schmiedeberg's Arch Pharmacol 317:228-232, 1981.

6. Colatsky TJ: Mechanism of action of lidocaine and quinidine on action potential duration in rabbit cardiac Purkinje fibers - an effect on steady state sodium currents? Circ Res 50:17-27, 1982.

7. Carmeliet E, Saikawa T: Shortening of the action potential and reduction of pacemaker activity by lidocaine, quinidine and procainamide in sheep cardiac Purkinje fibers - an effect of Na or K Currents? Circ Res 50:257-272, 1982.

8. DiFrancesco D, Ojeda C: Properties of the pacemaker current i_f in the sinoatrial node of the rabbit: A comparison with the current i_{K2} in Purkinje fibres. J Physiol 308: 331-351, 1980.

9. Dangman KH, Hoffman BF: Antiarrhythmic effects of ethmozin in cardiac Purkinje fibers: Suppression of automaticity and abolition of triggering. J Pharmacol Exp Ther 227: 578-586, 1983.

10. Ferrier GR, Saunders JH, Mendez C: A cellular mechanism for the generation of ventricular arrhythmias by acetylstrophanthidin. Circ Res 33:508-515, 1973.

11. Roden DM, Hoffman BF: Action potential prolongation and induction of abnormal automaticity by low quinidine concentrations in canine Purkinje fibers: relationship to potassium and cycle length. Circulation (in press).

12. Brachmann J, Scherlag BJ, Rosenshtraukh LV, Lazzara R: Bradycardia-dependent triggered activitity: relevance to drug-induced multiform ventricular tachycardia. Circulation 68:846-856, 1983.

13. Kass RS, Lederer WJ, Tsien RW, Weingart R: Role of calcium ions in transient inward currents and after contractions induced by strophanthidin in cardiac Purkinje fibres. J Physiol 281:187-208, 1978.

14. Dangman KH, Hoffman BF: In vivo and in vitro anti-arrhythmic and arrhythmogenic effects of N-acetyl procainamide. J Pharmacol Exp Ther 217:851-862, 1981.

15. El-Sherif N, Gough WB, Zeiller RH, Mehra R: Triggered ventricular rhythms in 1-day-old myocardial infarction in the dog. Circ Res 52:566-579, 1983.

16. Friedman PL, Stewart JR, Fenoglio JJ Jr, Wit AL: Survival of subendocardial Purkinje fibers after extensive myocardial infarction in dogs. In vitro and in vivo correlations. Circ Res 33:597-611, 1973.

17. Sasyniuk BI, Ogilvie RI: Antiarrhythmic drugs: Electrophysiological and pharmacokinetic considerations. Ann Rev Pharmacol Toxicol 15:131-155, 1975.

18. Eisner DA, Lederer, WJ, Sheu, SS: The role of intracellular sodium activity in the antiarrhythmic action of local anaesthetics in sheep Purkinje fibres. J Physiol 340: 239-257, 1983.

19. Gintant GA, Datyner NB, Cohen IS: Slow inactivation of a tetrodotoxin-sensitive current in canine cardiac Purkinje fibers. Biophys J 45:509-512, 1984.

20. Attwell D, Cohen I, Eisner D, Ohba M, Ojeda C: The steady state TTX-sensitive ('window') sodium current in cardiac Purkinje fibres. Pflugers Arch 379:137-142, 1979.

3

INITIAL EVALUATION OF NEW ANTIARRHYTHMIC AGENTS IN MAN: NORMAL VOLUNTEERS OR PATIENTS?

RAYMOND L. WOOSLEY, M.D., PH.D. AND JOHN A. OATES, M.D.

The pattern of development of new drugs has settled comfortably in place over the past 20 years. A traditional sequence of performing clinical testing of new chemical entities has evolved which attempts to maximize safety and protect the patient/volunteer. Under the usual scheme of things, following extensive toxicological evaluation in animals, a candidate drug is administered to apparently healthy human subjects, most often young adult males, in order to determine tolerance and establish an upper dose level beyond which undesirable side effects occur. There are of course, even now, exceptions to this trial of a potential new drug in healthy volunteers. Cytotoxic drugs, for instance, are not tested in healthy subjects, the obvious reason being that the benefit-to-risk ratio in the healthy subject is heavily weighted on the "risk" side. However, if one stops to consider for a moment, this ratio is invariably tilted in the risk direction, however slightly, when a healthy subject takes part in a Phase I study of any chemical compound. Tolerance, if not related to therapeutic effect, has

Supported by U.S. Public Health Service Grants 5MO1RR-95 and GM31304.
Dr. Oates is the Joe and Howard Wertham Professor of Investigative Medicine.

little practical import. Why expose healthy volunteers to a test compound until adverse effects occur, when the clinically effective dose may indeed be much smaller?

From a clinical viewpoint, how useful is the information gleaned from "traditional" Phase I studies? Doses to be employed in initial human testing are inferred from toxicological data obtained in animal studies. The relevance of animal data to drug disposition in humans has been debated (1). Often cited examples of discrepancies between species include the lack of teratogenicity of thalidomide in rats and the inability to reproduce the oculomucocutaneous syndrome associated with practolol. Given that wide species differences exist in the way drugs are handled in the body, much time, effort and expense may be invested administering doses to normal volunteers which, without any means of evaluating pharmacological effect, may be far from those eventually found useful in clinical practice. Furthermore, some data may be actually misleading when transferred to the patient population. For example, pharmacokinetic data usually gathered in healthy volunteers, often differs from that observed in patients.

Table 1 Antiarrhythmic Drug Elimination Half-Lives
(mean of reported values, hours)

	Normal Volunteers	Patients with Chronic Arrhytmias
Tocainide	11	14
N-acetylprocainamide	6.5	10
Mexiletine	10.5	12.5
Aprindine	22	48
Flecainide	14	20.3

As can be seen from Table 1, elimination half-life observed in normal volunteers may underestimate that in patients by as much as 100%. This discrepancy should come as no surprise. Differences in age, metabolism and concurrent disease processes all contribute to a different handling of the drug in the volunteer and patient populations. A further problem is that some drugs exhibit dose-dependent pharmacokinetics. For instance, hepatic metabolism of propranolol and phenytoin is saturable (2,3), as is disopyramide plasma protein binding (4). In such cases, without some knowledge of the therapeutic dose (which cannot be inferred from studies in healthy subjects), relevant plasma level information cannot be obtained. Pharmacokinetic constants calculated at a given dose level will not be applicable to a different dose.

The reasons for a reluctance to perform initial studies in patients are a complex interplay of medical, legal and emotional considerations. Arguments that patients should be spared the added burden of intensively monitored phase I studies, that patients may feel pressured to consent to a study in the belief that refusal might jeopardize the patient-physician relationship, that patients may litigate more readily should an untoward event occur, that the pharmacological effects of the investigational drug might be difficult to separate from those of patients' concomitant medication and/or disease processes have all been considered (5). Fortunately, in the case of antiarrhythmic agents, a patient population is available which is well suited to the initial testing of investigational compounds. Patients with stable cardiac rhythm disorders, such as high frequency ventricular ectopic depolarizations (VEDs) are often free from other illness including renal / hepatic disease. A single-blind placebo phase or a period of medication-free observation, provided of course that patients already receiving therapy can be safely weaned from their medication without inducing life-threatening or hemodynamically significant arrhythmias, is ethically justifiable since there is, as yet, no convincing evidence that suppression of asymptomatic VEDs provides prophylaxis against catastrophic cardiac events (6). Starting dosages may therefore be chosen well below those anticipated to be effective, thus providing maximum safety and minimal risk to the patient. Although there has recently been some question whether agents which are particularly effective in suppressing VEDs can provide adequate protection against more malignant arrhythmias (7), or may even precipitate electrical instability in patients with a history of ventricular tachycardia (VT) or fibrillation (8,9), in general there has been a reasonable correlation between data obtained in patients with chronic stable ventricular ectopy and that obtained later in patients with more serious arrhythmias. In addition, the group chosen for initial testing may be better able to tolerate any unforeseen hemodynamic or electrophysiologic effects of the test medication than other patient groups, such as those with compromised cardiac function or acute myocardial infarction, in whom the investigational compound will be subsequently used.

Obviously, the same care and caution exercised in healthy volunteer studies should be applied to studies in patients. Initial dosing should preferably be by slow intravenous infusion, since the vagaries of absorption after oral dosing are avoided and the infusion can be immediately discontinued should an adverse reaction occur. Table 2 suggests a dosing protocol which incorporates a wide safety margin.

Table 2. Suggested dosing protocol for initial phase I study

1. Initial low dosage (1/100th to 1/10th of predicted effective dose)
2. Gradual escalation
3. Limited dosing in an individual
4. Overlap in dosing during escalation in case of differences in responsiveness
5. Characterization of dose-response curve once an effect is identified

A period of monitoring to document arrhythmia stability is essential. This may require several days if the patient has been receiving other therapy or is unfamiliar with the new monitored environment. Once stable, a period of time, perhaps 24 hours, during which placebo is administered should be recorded as baseline, immediately preceding evaluation of the test drug. The importance of this early evaluation is shown in Figure 1 where the effect of hospitalization and placebo upon VED frequency is apparent.

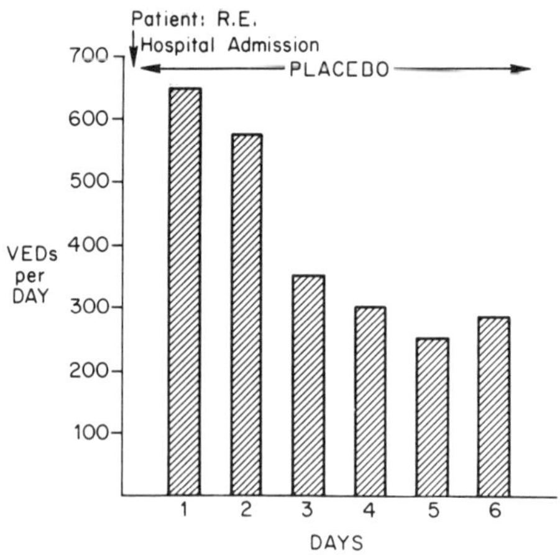

Figure 1. Effect of hospitalization and placebo on arrhythmia frequency.

Although a single dose is administered, patients should be observed for at least 24, and preferably 48 hours, after dosing. Since a decision to proceed to a higher dose rests on the results at preceding dose levels, rapid turnaround on analysis of ambulatory ECG tapes is obviously a necessity. Once a dose is defined which consistently produces an approximately 80% fall in VEDs a dose response curve can be better defined by using low, medium and high doses in 6-8 patients. Intensive monitoring of these patients can provide invaluable information on the relationship between dose, plasma concentration and effect, preliminary data on mechanisms of elimination and excretion, and the possible presence of active metabolites. If intravenous dosing has been used, studies of the oral preparation to demonstrate adequate bioavailability, and comparison of pharmacodynamic data between intravenous and oral routes of administration may be advisable at this point. If the pharmacologic effect after oral administration is much attenuated compared to intravenous dosing, then either impaired absorption or a significant first-pass effect should be suspected, with obvious implications for the continued development of the oral preparation; fecal excretion of unchanged drug will provide a clue as to what may be happening). If significant hepatic biotransformation is suspected, and larger oral doses are well tolerated (the metabolites have no or inconsequential pharmacologic activity), then the oral route may be viable; if larger oral doses are accompanied by unwanted effects (the metabolites have toxic, but little therapeutic effect, as in the case of lidocaine (10)), then the oral route may have to be abandoned. Thus, with perhaps a dozen or so patients exposed to the test drug under guarded conditions, sufficient data can be obtained to design and implement multiple dose studies with some degree of confidence in the dose range and dosing schedule to be employed.

A precautionary note is appropriate at this point. The patient with high frequency VEDs is still a "model" for the final population to be treated. Although effective dosages, pharmacokinetics and toxicity observed in this model may be similar, it is very likely they will still be different and these differences and these differences could have life-threatening implications. These studies can only provide guidelines for a conservative first approach to studies on other populations. The same dose-response (concentration-response) relationships will have to be defined again in these new patient populations. Cautious dosing and careful observation will be necessary to detect any implications from differences in pharmacokinetics, metabolism or inherent responsiveness to the drug.

Although patients with stable chronic ventricular ectopy provide an excellent population for preliminary drug evaluation, it must be kept in mind that not all antiarrhythmic agents are necessarily efficacious in suppressing VEDs, and yet may have marked activity against other arrhythmias such as recurrent VT or fibrillation. Some compounds may be clinically useful as antiarrhythmic agents because of actions to elevate the fibrillation threshold or selectively abolish only premature beats which initiate sustained tachyarrhythmias, but not alter overall ventricular ectopy (11,12). Obviously, such agents could not be adequately tested in patients with simple, high frequency VEDs. However, pharmacokinetic data and information on the tolerance of the drug in these patients can still prove invaluable when proceeding to protocols in patients with other arrhythmias against which efficacy might be demonstrated.

Fundamental to an understanding of the pharmacodynamics of any drug is some means of quantifying the compound in the body. Plasma level monitoring should play an integral part in the evaluation of a new agent, and a reliable and specific assay for an investigational drug should be made available as early as possible in its development. It should be clear whether the parent compound alone, or parent plus metabolites is being measured, and the results treated accordingly. In theory, although perhaps less often in practice, the drug concentration in plasma is in equilibrium with the drug concentration at the effector site. Thus, plasma level information is vital to our understanding of both the disposition of the drug in the body, and the relationship between plasma concentration and response, and thus to a preliminary definition of a therapeutic window for that drug. Cardiac agents, in general, have narrow therapeutic windows, that is, the range of plasma concentrations which are effective, but produce a minimal incidence of adverse effects, is small. Figure 2 illustrates the implications for choosing a dosing schedule when the therapeutic window is narrow, and when there is more leeway between the effective and toxic doses.

The therapeutic range also depends on the definition of efficacy. For example, in patients with recurrent non-sustained VT undergoing tocainide therapy, we found 90% of VT episodes could be suppressed at plasma concentrations of 2.8-10.0 (mean 6.0) mcg/mL, while 90% suppression of VEDs required 9.0-13.0 (mean 11.0) mcg/mL (13).

Another consideration in estimating the therapeutic range is the nature of the adverse effects which occur as the dose is increased. Mild side effects allow more flexibility in the upper limit of the therapeutic range, while the likelihood of severe adverse reactions, such as exacerbation of arrhythmias, dictates a conservative upper

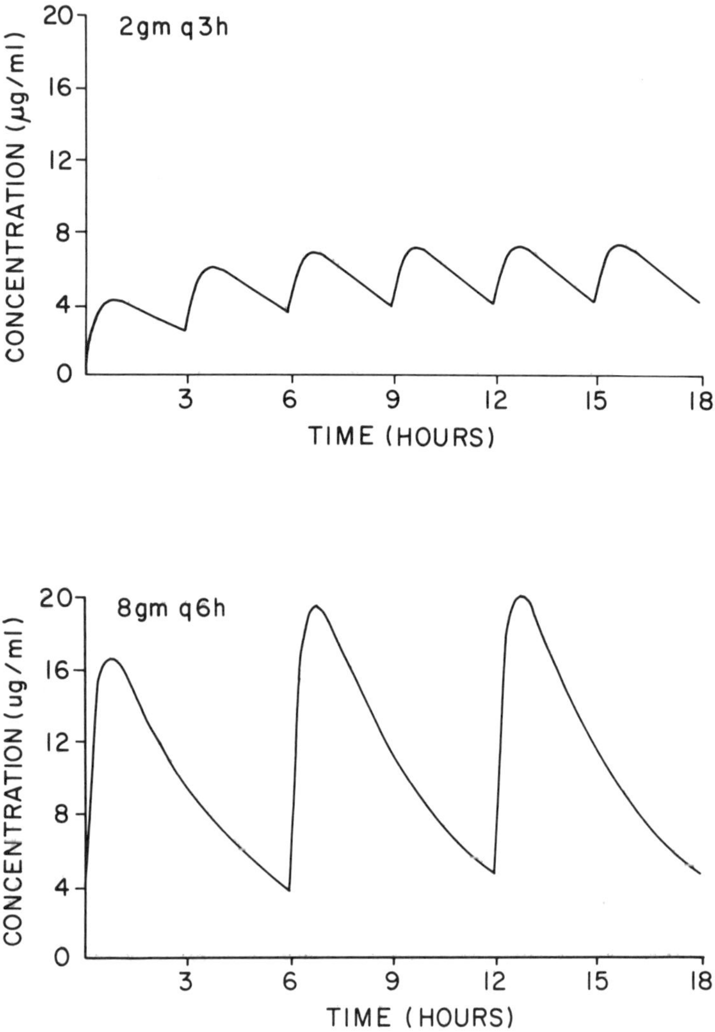

Figure 2. Relationship among the elimination half-life, the width of the therapeutic window and the frequency of dosing for a rapidly absorbed and eliminated drug (in this case procainamide). These computer simulations of plasma concentrations show that frequent dosing is required when the therapeutic window is narrow (4-8 mcg/mL; top panel) to avoid excessively high or low values. If the therapeutic window were wider (e.g.4-20 mcg/mL; bottom panel), larger doses could be given less frequently. Since elimination half-life (139 minutes) is not changed, the time to reach steady-state conditions is the same.

limit. Many adverse reactions are simply an extension of a drug's therapeutic effect. For example, an agent which prolongs atrioventricular conduction will likely also cause block at higher doses. Such considerations, together with pertinent patient observation in early studies (for example monitoring of ECG intervals) can therefore be invaluable both in anticipating certain adverse reactions and in providing insight into potential mechanisms for drug action. There will be instances, however, where toxic effects and/or therapeutic effects are unrelated to plasma concentration. For example, the antifibrillatory effects of bretylium correlate with drug concentration in the myocardium, rather than that in plasma (14); quinidine-induced torsades de pointes can occur at very low plasma concentrations of quinidine (15).

Plasma level monitoring, coupled with early comparison of pharmacologic effects after intravenous and oral dosing, also provides information on the possible presence of active metabolites. It should be appreciated that no special studies are necessary to infer the presence of active metabolites. If, after oral dosing, only a small amount of unchanged drug is recovered in the urine (fecal excretion of drug is usually low unless there are problems with absorption or significant enterohepatic circulation), and plasma levels of parent compound correlate poorly with pharmacologic effect with large interpatient variability in effective plasma concentrations, then the possibility of active metabolites should be considered. The realization that N-acetylprocainamide, an active metabolite of procainamide, is an antiarrhythmic in its own right, has been relatively recent (16). The fact that certain metabolites of quinidine might be arrhythmogenic and perhaps responsible for induction of torsades de pointes is only now being explored (17). However, we are learning to apply pharmacological principles in the testing of newer agents. Encainide and lorcainide are still investigational compounds, yet we already appreciate the contribution of active metabolites to their overall effect (Table 3).

Table 3. Changes in electrophysiologic parameters following intravenous and oral administration of encainide and lorcainide.

	Encainide		Lorcainide	
	Acute IV	Chronic oral	Acute IV	Chronic oral
AH interval	+8%	+42%	0	+23%
HV interval	+7%	+41%	+25%	+21%
atrial ERP	+6%	+21%	+ 2%	+13%
ventricular ERP	-1%	+13%	+ 3%	+17%

Data adapted from references 18, 19.

Since active metabolites often have pharmacological effects and pharmacokinetics distinct from their parent compound, their efficacy, toxicity, and time-course of accumulation should be evaluated alongside that of the parent compound. Awareness that the relative amounts of these metabolites may be genetically determined and extremely variable has only recently become appreciated. Our experience with encainide (20) and propafenone (21) demonstrates how polymorphic expression of drug metabolizing capacity can markedly influence pharmacokinetics.

In summary, by performing initial human studies in patients rather than volunteers much pharmacokinetic and pharmacodynamic information can be obtained which bears direct relevance to the population in which the drug will eventually be used. This is not a new idea; arguments to this end have been presented over the past ten years (2,22,23). However, it is often tempting to observe the status quo. Perhaps it is time to lend an ear to the arguments in favor of early studies in patients rather than healthy subjects. The investment of time and resources is precious, and can best be directed where the maximum benefit, to both patient and clinical investigator will accrue. We are becoming increasingly sophisticated in our technology; our approach to new drug development should not lag behind. Plasma level monitoring should be complementary to patient observation, since, together they provide invaluable insights into how we can best use a drug. We believe that optimum progress will be achieved by the systematic application of scientific method in the evaluation of new compounds. While the practice of medicine is an art, the development of a new drug is, more appropriately, an ever evolving science.

REFERENCES

1. George CF. The importance of clinical pharmacology in drug development. Clinical Science (60):247-250, 1981.

2. Wood AJJ, Carr K, Vestal RE, Belcher S, Wilkinson GR, Shand DG: Direct measurement of propranolol bioavailability during accumulation to steady state. Br J Clin Pharmacol (6):345-350, 1978.

3. Gugler R, Manion CV, Azarnoff DL: Phenytoin pharmacokinetics and bioavailability. Clin Pharmacol Ther (19):135-142, 1976.

4. Meffin PJ, Robert EW, Winkle RA, Harapat S, Peters FA and Harrison DC: The role of concentration-dependent plasma protein binding in disopyramide disposition. J. Pharmacokinetics and Biopharmaceutics, 7:29-46, 1979.

5. Weissman L: Multiple-dose Phase I trials - Normal volunteers or patients? One viewpoint. J Clin Pharmacol (21):385-387, 1981.

6. Winkle RA: Clinical efficacy of antiarrhythmic drugs in prevention of sudden coronary death. Ann N Y Acad Sci (382):247-255, 1982.

7. Chamberlain DA, Jewitt DE, Julian DG, Campbell RWF, Boyle DMcC, Shanks RG: Oral mexiletine in high-risk patients after myocardial infarction. Lancet (ii):1324-1327, 1980.

8. Winkle RA, Mason JW, Griffin, JC, Rose D: Malignant ventricular tachyarrhythmias associated with the use of encainide. Am Heart J (102):857-864, 1981.

9. Oetgen WJ, Tibbits PA, Abt MED, Goldstein RE: Clinical and electrophysiologic assessment of oral flecainide acetate for recurrent ventricular tachycardia: evidence for exacerbation of electrical instability. Am J Cardiol (52):746-751, 1983.

10. Narang PK, Crouthamel WG, Carliner NH, Fisher ML: Lidocaine and its active metabolites. Clin Pharmacol Ther (25):654-662, 1978.

11. Sanna G, Arcidiacono R: Chemical ventricular defibrillation of the human heart with bretylium tosylate. Am J Cardiol (32):982-987, 1973.

12. Waxman HL, Groh WC, Marchlinski FE, Buxton AE, Sadowski LM, Horowitz LN, Josephson ME, Kastor JA: Amiodarone for control of sustained ventricular tachyarrhythmias: Clinical and electrophysiologic effects in 51 patients: Am J Cardiol (50):1066-1074, 1982.

13. Roden DM, Reele SB, Higgins SB, Carr RK, Smith RF, Oates JA, Woosley RL: Tocainide therapy for refractory ventricular arrhythmias. Am Heart J (100):15-22, 1980.

14. Anderson JL, Patterson E, Conlon M, Pasyk S, Pitt B, Lucchesi BR: Kinetics of antifibrillatory effects of bretylium: correlation with myocardial drug concentrations. Am J Cardiol (46):583-589, 1980.

15. Roden DM, Woosley RL, Bostick D, Bernard Y, Primm RK: Quinidine-induced long QT syndrome: incidence and presenting features. Circulation (68 Suppl III):III-276, 1983.

16. Roden DM, Reele SB, Higgins SB, Wilkinson GR, Smith RF, Oates JA, Woosley RL: Antiarrhythmic efficacy, pharmacokinetics and safety of N-acetylprocainamide in human subjects: comparison with procainamide. Am J Cardiol (46):463-468, 1980.

17. Drayer DE, Lowenthal DT, Restivo KM, Schwartz A, Cook CE, Reidenberg MM: Staedy-state serum levels of quinidine and active metabolites in cardiac patients with varying degrees of renal function. Clin Pharmacol Ther (24):31-39, 1978.

18. Jackman WM, Zipes DP, Naccarelli GV, Rinkenberger RL, Heger JJ, Prystowsky EN: Electrophysiology of oral encainide. Am J Cardiol (49):1270-1278, 1982.

19. Echt DS, Mitchell LB, Kates RE, Winkle RA: Comparison of the electrophysiologic effects of intravenous and oral lorcainide in patients with recurrent ventricular tachycardia. Circulation (68):392-399, 1983.

20. Woosley RL, Roden DM, Duff HJ, Carey EL, Wood AJJ, Wilkinson GR: Co-inheritance of deficient oxidative metabolism of encainide and debrisoquine. Clin Res (29):501A, 1981.

21. Siddoway LA, McAllister CB, Wang T, Bergstrand RH, Roden DM, Wilkinson GR, Woosley RL: Polymorphic oxidative metabolism of propafenone in man. Circulation (68, Suppl III): III-64, 1983.

22. Azarnoff DL: Physiologic factors in selecting human volunteers for drug studies. Clin Pharmacol Ther (13):796-802, 1972.

23. Oates JA: A scientific rationale for choosing patients rather than normal subjects for phase I studies. Clin Pharmacol Ther (13):809-811, 1972.

4

IS THERE A RATIONAL BASIS FOR THE MODIFIED CLASSIFICATION OF ANTIARRHYTHMIC DRUGS?

DONALD. C. HARRISON, M.D.

1. INTRODUCTION

Over the last 25 years a number of attempts have been made to classify antiarrhythmic drugs according to their intracellular electrophysiologic properties, their electrophysiologic properties in the intact heart (where studies could determine thresholds of conduction), and their clinical use in specific syndromes. Most of these attempts did not increase our knowledge of antiarrhythmic drugs or their clinical application. Since quinidine, introduced in the 1930's, was the only antiarrhythmic drug available until procainamide was discovered in the mid-1950's, no classification was necessary at first. During the 1960's, basic electrophysiologists studied the activity of several compounds with antiarrhythmic properties on isolated cells from the heart. Miles Vaughn Williams presented a classification scheme for antiarrhythmic drugs in 1971 at an international meeting of cardiologists and electrophysiologists in Elsinor, Denmark.[1] This scheme has been widely adopted, and physicians have tried to fit new drugs into it. Vaughn Williams' classification, and prototypical drugs for each class, are illustrated in Table 1.

Table 1. Vaughn-Williams Classification.

Class	Actions	Prototype Agent
I	Inhibits Na^+ transport Reduces dV/dt of Phase 0	Quinidine
II	Beta blockers	Propranolol
III	Prolongs repolarization Alters membrane response	Amiodarone
IV	Calcium channel block	Verapamil

In 1980 I modified Class I of the Vaughn-Williams classification to include several new antiarrhythmic drugs, thus making it more useful to the basic scientist and clinician.[2] This review defends the proposed modified classification.

2. PROPOSED CLASSIFICATION OF ANTIARRHYTHMIC DRUGS

At an international symposium on antiarrhythmic drugs held at Stanford University in 1979, I proposed a new breakdown of Vaughn-Williams Class I agents. Several new antiarrhythmic drugs had become available for clinical use, and basic cellular studies had provided sufficient data to classify them into Class I subgroups.[3,4] This classification is outlined in Table 2.

Table 2. Harrison Modification of Class I Agents.

Class	Actions	Agents
IA	Slows dV/dt of Phase 0 Moderate prolongation of repolarization Prolongs PR, QRS, and QT intervals	Quinidine Procainamide Disopyramide Cibenzoline
IB	Limited effect on dV/dt of Phase 0 Shortens repolarization Shortens QT in clinical doses Elevates fibrillation thresholds	Lidocaine Tocainide Mexiletine Ethmozine
IC	Markedly slows dV/dt Little effect on repolarization Markedly prolongs PR and QRS on ECG	Encainide Lorcainide Flecainide Propafenone Indecainide

The new classification of antiarrhythmic drugs is based on studies in isolated cardiac tissue, including muscle and His-Purkinje tissue. By using data from sophisticated electrophysiologic studies, intervals can be measured to the nearest millisecond, and the effects of compounds on atrial and ventricular refractoriness, fibrillation thresholds, and conduction properties can be described in detail. Specific data are also provided on the effects of new agents on the electrocardiogram used clinically to determine the electrophysiologic properties of drugs.[3,4]

New drugs studied in the 1970's and 80's include flecainide, lorcainide, encainide, propafenone, ethmozine, tocainide, mexiletine, indecainide, and cibenzoline. While all electrophysiologic and clinical data are not available for these new agents, sufficient studies have been performed in basic and clinical electrophysiologic laboratories to classify them.[3,4] As new drugs become available, their electrophysiologic properties in cells and in man will determine their classification. This scheme is used to discuss indications for the use of specific drugs for specific arrhythmias in the following papers.

3. CLINICAL USE OF THE MODIFIED CLASSIFICATION SCHEME

Any classification scheme which is to be relevant to the clinician must help him make clinical decisions for prescribing drugs for patients with arrhythmias. Knowing the electrophysiologic properties for each drug permits a more rational basis for their use in man. For example, Class IA drugs are hazardous for patients having arrythmias with a prolonged Q-T interval, since they markedly prolong repolarization and the Q-T interval. Using this scheme, the cardiologist can also predict the most effective drug for a particular clinical arrhythmia. It was predicted, for instance, that drugs with profound effects on conduction velocity in the His-Purkinje system would be useful for patients with well-established bypass tracts and the Wolff-Parkinson-White syndrome; encainide has now been proven of major importance in preventing the recurrent re-entry arrhythmias of that syndrome.

However, the modified classification scheme can most usefully aid the clinician by allowing him to develop rational combinations of drugs for patients with resistant and intractable arrhythmias. For example, combining the properties of one drug from Class 1A and one from Class 1B can achieve maximal effects on fibrillation thresholds while keeping prolongation of repolarization to a minimum. Also, since many patients cannot tolerate high doses of certain antiarrhythmic drugs, lower doses of drugs from different subgroups let patients benefit from their additive electrophysiologic properties without producing major undesirable side effects. A third use of this classification is to warn the clinician not to combine Class 1A with Class IC drugs, because both have additive electrophysiologic properties which may sometimes be helpful and sometimes harmful. Instead, either a Class IA or a Class IC drug combined with a Class IB drug would provide beneficial electrophysiologic effects for patients with severe and life-threatening arrhythmias.

4. FUTURE DIRECTIONS

While new drugs can be easily added to the classification scheme presented above, problems can be expected as we learn more about the pharmacology and physiology of antiarrhythmic drugs. Some of these are:

4.1 Metabolites of anti-arrhythmic drugs

Several antiarrhythmic drugs have active metabolites which occur after hepatic breakdown of the parent compound in man (Table 3) and which can be studied electrophysiologically and clinically.[5,6,7] Some have electrophysiologic profiles similar to their parent compounds, but may have different clinical actions in man. In addition, they clearly have altered pharmacologic properties, and may accumulate to produce some of the pro-arrhythmic effects noted with the Class IC drugs. Also, some metabolites have qualitative differences from their parent compounds and may belong in a different classification.

Table 3. Antiarrhythmic Drugs with Active Metabolites.

Parent Drug	Metabolite	Activity
Quinidine	4-hydroxyquinidine	Equal to quinidine
Procainamide	N-acetyl-procainamide	25% of procainamide
Lidocaine	1. Monoethylglycinexylidide (MEG_X)	30% of lidocaine
	2. Glycinexylidide (G_X)	50% of lidocaine
Encainide	1. 0-demethylencainide	Greater than encainide
	2. 3-methoxy 0-demethylencainide	Equal to encainide
Lorcainide	Norlorcainide	Equal to lorcainide

4.2 New agents

As the pharmacologic industry increasingly develops molecular structures with persistent effects, antiarrhythmic drugs will be coupled to drugs with other properties, such as diuretics, anti-hypertensives, and calcium channel blocking agents. How will such drugs fit in the modified classification scheme?

4.3 New electrophysiologic properties which may have antiarrhythmic action

Some compounds produce profound changes in the electrophysiology of isolated myocardium or His-Purkinje tissue but have little effect in clinically-tolerated doses on arrhythmias in man. How such agents can be classified is yet to be determined.

5. SUMMARY

In summary, I believe the modified classification scheme for antiarrhythmic drugs in Vaughn Williams' Class I has a rational basis for the clinician. It enables him to predict the appropriate sub-class of drugs for patients with particular arrhythmias and to determine combined drug therapy for patients with life-threatening cardiac arrhythmias. This subclassification has effectively communicated clinical ideas on antiarrhythmic drugs over the past four years and should continue to prove useful in the future.

REFERENCES

1. Vaughn Williams EM: Classification of antiarrhythmic drugs. In: Sandøe E, Flensted-Jensen E, Olesen K, eds. Cardiac Arrhythmias. Södertalje, Sweden, Ad Astra, 1970, pp. 449-473.
2. Harrison DC, Winkle RA, Sami M, Mason JW: Encainide: a new and potent antiarrhythmic agent. In: Harrison DC, ed. Cardiac Arrhythmias: A decade of progress. Boston, G. K. Hall Medical Publishers, 1981, pp. 315-330.
3. Keefe DLD, Kates RE, Harrison DC: New antiarrhythmic drugs: their place in therapy. Drugs 22:363-400,1981.
4. Harrison DC, ed.: Symposium on perspectives on the treatment of ventricular arrhythmias. Am J Cardiol 52:1C-2C, 1983.
5. Woosley RL, Roden DM: Importance of metabolites in antiarrhythmic therapy. Am J Cardiol 52:3C-7C, 1983.
6. Harrison DC, Irwin WG: New concepts for antiarrhythmic drugs in the 1980's. Arch Mal Coeur (special):73-77, 1981.
7. Kates RE, Woosley RL, Harrison DC: Clinical importance of metabolites of antiarrhythmic drugs (editorial). Am J Cardiol 53:248-251, 1984.

PANEL ON BASIC CONSIDERATIONS

Dr. Reder: Dr. Hoffman, would you speculate on the possible importance of the *in vitro* phenomenon of triggered activity and clinical ventricular arrhythmias?

Dr. Hoffman: I think that some arrhythmias of the human heart particularly those caused by excessive doses of digitalis almost certainly result from triggering of delayed after depolarizations. I think that the type of arrhythmia that people classify as Torsade in patients that are taking certain classes of antiarrhythmic drugs such as quinidine or amiodarone where the onset of the arrhythmia follows a long cycle or occurs during bradycardia and where the arrhythmia can be completely suppressed by pacing the heart at a rapid rate or giving isoproterenol, certainly are triggered arrhythmias due to early after-depolarizations. I firmly believe that both mechanisms can and do cause specific types of arrhythmias in human hearts. I don't know how to make a general statement about all of the other arrhythmias that one can record from the human heart which might be due to after-depolarizations. There are some tests being developed to improve the diagnosis of triggered rhythms. The tests are still not completely adequate. If an arrhythmia can be suppressed, by overdrive, I feel it is probably not triggered and if an arrhythmia follows Waldo's entrainment rules, I think it probably is not triggered. All I would really say is that triggering due to either early or delayed after-depolarizations does cause arrhythmias in the human heart and it is up to the cardiologist to figure out which is which.

Dr. Harrison: I would agree that there are arrhythmias in man probably caused by this mechanism. Our problem has been that we have not had a good way to study them without going through some of the tests that provide us with the answer to the rules that Dr. Hoffman mentioned. There are a lot of new techniques presently being developed that at least potentially are going to allow one to look at this problem. Some of the computer based techniques with signal-averaging may well provide a handle to look at this question, and if that is true then I think we may be able to do something with the whole concept of after-potentials and the triggering of arrhythmias.

Dr. Moore: I think that electrophysiologists are a little bit like surgeons. When rheumatic heart disease declined, cardiac surgeons developed the CAVBG procedure when it looked as if clinicians could distinguish arrhythmias due to reentry from automaticity by programmed stimulation, then electrophysiologists introduced the complications of membrane oscillations, after-depolarizations, triggered automaticity, and slow channel potentials. As Dr. Hoffman and Dr. Harrison have said, these may be an important cause of arrhythmias. Our lab has found in animal as well as human hearts that both reentry and altered automaticity can occur in the same heart. Since different drugs have different actions on both of those mechanisms, it seems little wonder that tachyarrhythmias can be difficult to treat. I would like to ask Dr. Woosley and Dr. Campbell to comment on the importance of triggered activity and after-depolarizations in human arrhythmias.

Dr. Woosley: I think that clearly anything other than what Dr. Hoffman says is going to be speculation. To me it has been interesting that there seems to be a group of patients that come in to see physicians who have a waveform of their ventricular tachycardia that looks an awful lot like triggered activity. These people have repetitive monomorphic ventricular tachycardia that seems to be related to catecholamines, stress or physical activity. It seems to warm up and slows down before it breaks and the arrhythmias have a lot of characteristics that makes you want to select these people out and study them more intensively.

Dr. Campbell: I think we have a basic problem because we have such crude techniques in clinical situations for determining arrhythmia mechanisms. To actually determine what is going on at the cellular level from the surface electrocardiogram or from a few selected endocardial or epicardial sites per-operatively is very difficult. I have made a great play that we may inadequately define arrhythmias that we see. There are many characteristics of the electrocardiogram that we fail to analyze. We need to detail as many of the aspects of arrhythmias as may be relevant to determining mechanisms including the coupling intervals, the morphologies, and the interectopic intervals of a variety of tachyarrhythmias. I think that Torsade de Pointes underscores the importance of this and probably reflects a unique electrophysiological situation. We need to tease out the surface ECG appearances a little more and get back to fundamental mechanisms.

Dr. Morganroth: I have a question on a different subject. If one is going to do phase one studies, looking at early efficacy in man for ventricular arrhythmias and antiarrhythmic agents, should one use patients with stable high frequency arrhythmias as Dr. Woosley outlined. I am concerned about primarily using that group of patients with benign vs. potentially lethal arrhythmias. In the vast majority of patients are very variable and usually of relatively low frequency, between 10 and 100 per hour over a 24 hour period.

Is there any data that correlates how drugs work in high frequency patients that are relatively less variable than in variable low frequency patients? I think more commonly in clinical practice we deal with the patients that are more variable. Should these patients be included in the phase 1 studies in addition?

Dr. Woosley: I don't know of or have any nice tabular data. I can only relay some anecdotes. All of the drugs that have been effective in stable PVC patients have found clinical utility in patients with malignant arrhythmias to date. There may have been some drugs that didn't have utility in the stable PVC patients and thus never got to the malignant population, so we may have some false negatives in our studies that we will never know about. We found with Tocainide that it was very effective in the early people that we placed into our dose-response studies to look for pharmacologic effect; a 75% response rate when we selected people who had some historical response to lidocaine. When we moved into the population that you are talking about, the 10 PVC/hr, the stable outpatients with PVC's, we had about a 20% response rate. When the drug was put into malignant arrhythmias the response rate fell to 10 to 20% at most. I think that trend is there probably for all drugs, but it is not well established because the ability to make correlations between dose and effect is difficult. Concentration effect are greatest at one end of the spectrum and least at the other, so the correlations are going to be tough to come by. I think that the trend is if a drug works in one PVC population, it will probably work in the others. However, the dose response may be different and the percent response may be very different.

Dr. Harrison: Although I agree with Dr. Woosley's concept of studying this group of patients, I think it would be very difficult to predict from this group of patients what the response is going to be in another group of patients, due to a different mechanism for the arrhythmia in some of these patients. When people with relative high frequency ectopic activity and some degree of stability respond, it is generally my experience that the patients with lower frequency respond quite well too. This is particularly true if you stay in the homogeneous etiology such as coronary artery disease. If you switch to cardiomyopathies, prolapsed mitral valve, and everything else, then I don't think that is true. It is the other end of the spectrum where I think you see difficulties. In the patients with life-threatening serious ventricular tachycardia or people who have been defibrillated with ventricular fibrillation, the frequency of response then drops quite markedly with any of the drugs that we have studied.

Dr. Woosley: Another point I should mention is that in our experience, the thing that you don't see in these early studies, that you would like to see is pro-arrhythmogenicity. That usually comes out later in the more malignant arrhythmias for a number of reasons. One is in some cases, we didn't recognize the marked differences in pharmcokinetics between the two populations. I think that a lot of the people who developed problems with Flecainide were given overdoses, and that in many people who developed arrhythmogenicity with Encainide, the active metabolite wasn't well recognized. The time to steady state certainly wasn't recognized. Now we know with Encainide and probably with Flecainide, that it takes 3 to 4 days before steady state effects are seen. It has been difficult to pick up arrhythmogenicity in these early studies, but it sure comes on with a bang later.

Dr. Campbell: I am concerned that using this PVC patient group is profiling drug development and drug investigation towards this one subset of the population with PVC's. I think that the major public health problem relates primarily to the prevention or management of ventricular fibrillation and sudden death. Drugs such as the beta-adrenoreceptor blocking drugs which may prevent ventricular fibrillation probably might not be adequately tested in the models that you are suggesting.

Dr. Woosley: I don't propose that these are adequate tests of any drug. I am simply proposing them as the first step to find the dosage range that might be worth evaluating in subsequent studies.

Dr. Campbell: But do you think that you would have passed many of the beta-adrenoreceptor blocking drugs through your investigational protocol and approved them.

Dr. Woosley: I think that is a different type of drug requiring a different model for phase 1 and 2 testing. The problem is that if you have an antiarrhythmic drug and you ask me to put it into a sudden death study, I'll ask you what dose.

Dr. Harrison: In regards to beta-adrenergic blocking drug actions and sudden death, there are a lot of other explanations of what is going on beyond antiarrhythmic actions. In fact I think beta blockers would have missed the screen that Dr. Woosley is talking about. I don't know how you would predict the fact that they would prevent sudden death.

Dr. Campbell: I totally agree. The problem obviously is, how do we detect antifibrillatory drugs. I am concerned that most of the industry's models for determining antiarrhythmic activity still relate to models that are very similar to the clinical ones that you have proposed. We have not developed the screening methods of identifying drugs that may have powerful anti-VF actions.

Dr. Woosley: I think that VF and sudden death is an over-used term, but it is multifactorial and it may require a multitude of approaches and attacks. The PVC suppression hypothesis is still being tested. Antiarrhythmic drugs have the best chance of providing a mechanism to test that hypothesis.

Dr. Chun: I have reviewed a few antiarrhythmic drugs so I would like to make a comment about the anti-PVC effect of the drug studies. I think that Dr. Woosley's statement is correct. When you evaluate the anti-PVC effect by the PVC numbers, we found that the patients who had higher PVC numbers in the baseline were easier to show anti-PVC effect compared to those who had a few PVC per hour for evaluation of drug effect. We evaluated the organic cardiac condition with different PVC populations and found that there was no difference between coronary heart disease and other cardiac diseases. We do not know what is going to be the effect for the prevention of sudden death; that is another problem. I have one comment about Dr. Woosley's phase 2 studies regarding stepping up the dose to find the effective drug dose. In retrospective studies, we found that the so-called effective dose protocol as determined by step-wise increasing the dose caused more toxicity. This is because 2 to 3 days of step-up dosing to find the so called effective dose, results in the dose being too high in long multiple studies. Of course this may be due to the active metabolite(s). I am sure that there are active metabolites with activity which we do not identify at this time. For instance, Flecainide is not supposed to have an active metabolite. When a very high frequency of reactions was found and the dose was cut down to half, the effectiveness was still there but the side effect were far less than the so-called effective dose. My recommendation is that after the step-wise increased dose to establish the so-called effective dose is determined, then the next stage is to determine the lowest effective dose. I think that would be better for evaluation in phase 3 and phase 4 studies.

Dr. Woosley: I agree with Dr. Chun entirely. I think that the assumption for any study design is that we have a good understanding of the pharmacokinetics from phase 1 and 2. In the case of Encainide we certainly did not know about metabolites and that led us astray. I think with Flecainide that it is a very complex issue. Maybe it is just that we don't have good pharmacokinetics allowing us to predict the duration of an effect. I think it is a beginning point. The other phase that I did not discuss is a re-institution phase where we follow the patients chronically and try to take them to the lowest effective dose. That is not part of our protocol, but it is certainly a part of our overall drug evaluation.

Dr. Harrison: In terms of evaluation of PVC frequency, there are statistical techniques and other ways to define populations. Dr. Sami worked with us using an analysis of variance. I think you can study PVC's in patients with lower frequencies than we have in the past depending on how you define your population, and how you characterize it before you start your studies. A major question is the etiology of the PVC. I still believe that there is a considerable difference in the response in various patient groups and particularly in patients with prolapsed mitral valve. I find them very difficult to predict and very difficult to deal with. I don't think they generally should be put into the same study protocols as patients with coronary heart disease.

Dr. Woosley: I think they can go in there but that they have to be characterized.

Dr. Temple: I was struck by Dr. Campbell's suggestion that perhaps these PVC models cause us to miss drugs with antifibrillatory activity. I think the problem is in the other area. We don't know how to test drugs that are hypothetically antifibrillatory in a reasonably efficient way. Every such study requires a huge mortality study and we are not good at doing those frequently or with adequate numbers.

Dr. Boal: One of the problems that I see that has not yet been addressed is that pharmacodynamics are very frequently measured in normal healthy volunteers and these are not age matched to the patient populations that we treat. The best example of this is digoxin which was originally tested in normal healthy volunteers. When such pharmacokinetics were applied to 80 year old patients they just absolutely didn't apply. That might be one argument in favor of using a group closer to the patient population that we are ultimately going to treat.

Dr. Harrison: I agree that we have got to study the pharmacokinetics in patient populations where they will be used, in the elderly, in the patients with congestive heart failure, patients with renal and liver disease. I don't think any of us would imply anything else.

Dr. Woosley: Maybe someone else can come up with a reason to study the kinetics in normal volunteers. I can't think of one.

Dr. Lin: Dr. Woosley, could you give us a definition of what is a young potentially catastrophic stable ventricular arrhythmic patient?

Dr. Woosley: Someone with the three to four beats of ventricular tachycardia that is asymptomatic who is willing to come into the hospital to spend three weeks of his time or her time in a drug trial. These people are hard to find. They may be very different from the final patient population, but they are out there. They can be found. Patients with prolapsed mitral valve, again, asymptomatic. Patients with a cardiomyopathy that has not deteriorated to the point that their hepatic or renal function is abnormal. We like patients who have greater than 1 PVC/minute for the earliest studies. When we get down to 10 per hour as Dr. Chun pointed out, it sometimes becomes difficult in the individual to tell if the drug is working. Dr. Harrison and I have shown that for a group of people you can demonstrated a pharmacologic effect in patients with 10/hour, but in an individual, it becomes very difficult to know if a dose has done anything to PVC frequency when you are that low.

Dr. Temple: The reason you do studies in relatively normal people is because if you do them in complicated non-normal people, you won't know what your findings mean. If you take an old, sick population and get kinetic information, you won't know how the patient's renal function, how the patient's cardiac function, how the patient's body mass, etc. have modified what the normal kinetics are. You need to have some kind of baseline in reasonably normal people. That doesn't mean they have to be non-hypertensive people or people who don't have any VPB's but you want them to be relatively normal from which you can then move on to people with reasonably well defined impairments or changes in body habits or whatever, to see what influence those things have. Doing them in otherwise normal people who have VPB's is perfectly normal.

II. STATUS OF SPECIFIC ANTIARRHYTHMIC AGENTS

5

PROCAINAMIDE, QUINIDINE, DISOPYRAMIDE, CIBENZOLINE, PIRMENOL - EFFICACY IN THE TREATMENT OF VENTRICULAR ARRHYTHMIAS: CURRENT STATUS AND CONTROVERSIES

CRAIG M. PRATT, M.D.

I. Introduction

In the review of the Type IA antiarrhythmic drugs, I will focus on the present knowledge regarding relative efficacy and toxicity in treating ventricular rhythm disturbances, and highlight a few areas of recent investigational interest including the antiarrhythmic efficacy of NAPA, bioavailability of sustained release preparations, as well as the issue of "proarrhythmia" with this group of medications. The assets and liabilities of the present classification system of antiarrhythmic agents has already been reviewed in this symposium. Suffice it to say that these drugs share certain electrophysiologic properties including their ability to decrease dV/dT of phase 0, producing conduction slowing, prolonging action potential duration and refractoriness, depressing membrane excitability, membrane responsiveness, and automaticity in normal Purkinje fibers.[1] The current status of cibenzoline and pirmenol, two investigational agents for the treatment of ventricular rhythm disturbances, are also reviewed. Both these drugs share many electrophysiologic properties with the Type IA drugs.

II. Procainamide:

Efficacy in the Treatment of Ventricular Arrhythmias.

Procainamide has been used in the United States for the treatment of ventricular arrhythmias for over 35 years.[2] Multiple case series documented the ability of procainamide to suppress ventricular premature depolarizations, including ventricular tachycardia, even in the setting of acute myocardial infarction.[2,3,4] This included the observation that the drug was effective at plasma concentrations of 4 8 µg/ml, although occasional patients required plasma levels double that for arrhythmia suppression. It was also noted that toxic manifestations occurred more commonly in patients with plasma concentrations of greater than 16 µg/ml.[4]

The major deficit with early papers is the lack of quantitative ambulatory monitoring. Recent studies include the evaluation of 17 patients with frequent ventricular premature depolarizations by Winkle et al[5], who reported that two thirds of the patients had an $\geq$ 90% reduction in VPDs. Unfortunately, gastrointestinal and lupus-like symptoms were severe, and 7 of the 17 patients were unable to complete treatment with procainamide at a dose of 3 g/day. In his review of 18 patients with ventricular arrhythmias, Myerberg reported the variable plasma level of procainamide required to suppress ventricular rhythm disturbances, depending upon the specific grade of ventricular arrhythmia and extent of organic heart disease.[6] For instance, in acute myocardial infarction a plasma concentration of 5 μg was adequate to achieve a 85% or greater suppression of VPD's, whereas a plasma concentration of nearly twice that (9.3 μg/ml) was required for an 85% suppression of VPD's in the setting of chronic ischemic heart disease.[6] In patients with sustained ventricular tachycardia, prevention of recurrence of sustained VT was achieved with a plasma procainamide concentration of 9.1 μg/ml, whereas a higher plasma concentration of procainamide (14.9 μg/ml) was required for an 85% suppression of VPD's in the same patients. These important observations highlighted the inadequacy of a simplistic "therapeutic plasma level" to define efficacy in all patient populations.

Programmed Electrical Stimulation - What Has it Added to Our Knowledge of Procainamide?

Procainamide has been investigated in patients with nonsustained and sustained ventricular tachycardia by electrophysiologic testing. In his report of 47 patients with spontaneous and inducible nonsustained ventricular tachycardia, Buxton used procainamide in 29 patients, resulting in prevention of induction of nonsustained ventricular tachycardia in 16 of 29. The mean serum procainamide level of 10.2 μg/ml was not significantly different from the level in patients who were nonresponders (10.5 μg/ml). An important observation, confirmed by multiple other investigators, was that patients with underlying heart disease were more refractory to therapy than those patients with no cardiac diagnosis and normal ventricular function.[7] In a separate report of 35 patients referred for the electrophysiologic evaluation of recurrent ventricular tachycardia, Greenspan et al [8] reported on 16 patients in whom procainamide was determined by programmed ventricular stimulation and serial drug testing to be the most effective antiarrhythmic agent. In the majority of these cases the clinical arrhythmia involved was sustained ventricular tachycardia. In these "responders" to procainamide, these investigators reported a high dose of oral procainamide was necessary (mean dose 6 g/day) to achieve a mean plasma level of 13 μg/ml. Although the authors concluded that high dose procainamide was effective in the treatment of recurrent inducible ventricular tachycardia, 25% developed lupus side effects.[8]

The same group reported on 126 patients with inducible sustained ventricular tachycardia during acute electrophysiologic testing. Of practical importance was the fact that intravenous procainamide, when found to be effective (30 patients) predicted a patient population in whom at least one other effective alternative drug regimen was found (83% of patients). This was in sharp contrast to the procainamide unresponsive patients in whom effective alternative drug regimens were found in only 9% of patients. There was no predictive value of the plasma procainamide level as to which patients were responders.[9]

Is There Evidence that Procainamide Prevents Sudden Cardiac Death?

Although large scale clinical trials are lacking, Myerburg et al reported on 16 patients who had previous episodes of documented sudden cardiac death. These patients had been placed on procainamide at a dose of 2-5 g/day with a goal of achieving Procainamide levels between 4-8 μg/ml. Eight of these patients had a recurrent episode of sudden cardiac death in the one year followup period and all were characterized by unstable plasma levels of procainamide, usually subtherapeutic, during that interval.

In summary, Procainamide seems to be an adequate suppressant of VPD's (an average 80% suppression of VPD's reviewing all studies) and appears to be successful in at least half the patients who manifest sustained ventricular tachycardia. Its efficacy in preventing sudden cardiac death is not established except for small case series. In patients with sustained ventricular tachycardia and/or sudden cardiac death the doses of procainamide required appear to be large ($\geq$6 g/day) with concomitant serum procainamide levels above 13 μg/ml. Toxicity to procainamide is common, with gastrointestinal disturbances and lupus-like symptoms occurring in at least 25% of all patients. This appears to be somewhat dose related. The issue of "proarrhythmia" will be dealt with in a later section.

Do Sustained Release Procainamide Preparations Provide Adequate Serum Levels?

In a study of 7 cardiac patients who were given equivalent daily doses of bot procainamide as well as a sustained release preparation, Vlasses[11] demons trated a serum trough procainamide level of 20.3 μg/ml compared to 18.7 μg/m with the sustained release preparation. Equivalent values for NAPA were 25. μg/ml and 23.4 μg/ml respectively. Thus, at equal total daily doses, troug plasma NAPA levels were similar. Giordina[12] and colleagues have examined th plasma concentration as well as antiarrhythmic efficacy of sustained releas procainamide in 33 patients with heart disease. Using a mean dose of sustaine release procainamide of 4.8±1.5 g/day, they achieved an $\geq$ 75% reduction i VPD's in 25 of the 33 patients. There was a > 95% reduction in both paire forms and runs of ventricular tachycardia. This was achieved with a mea trough procainamide concentration of 6.8 μg/ml and concomitant NAPA concen tration of 8.7 μg/ml. In the vast majority of these patients, the troug plasma procainamide and NAPA concentrations were $\geq$ 50% of the peak plasm concentration. Of interest was the fact that these investigators reporte minor side effects in approximately one half of the patients, in whom ther was a higher mean plasma procainamide and NAPA concentration. These investiga tors and others have confirmed the fact that procainamide sustained releas preparations are capable of effectively doubling the half-life of conventiona Procainamide to approximately 6 to 7 hours.[13,14]

III. N-acetyl Procainamide (NAPA) - How Effective Is It?

Although procainamide is moderately effective in suppressing prematur depolarizations, nonsustained ventricular tachycardia, and selected cases o sustained ventricular tachycardia, it possesses multiple characteristic making it undesirable as an ideal antiarrhythmic drug. Its short eliminatio half-life necessitates frequent dosing. Also, the frequent occurrence o lupus-like side effects, especially in patients taking high doses o procainamide decreases its desirability. The occurrence of torsade de pointe ventricular tachycardia will be discussed subsequently.

Therefore, the potential efficacy and advantages of NAPA were investigated b Winkle and his associates, who studied 11 patients with frequent VPD's. spectrum of doses from 500 mg to 2500 mg every 8 hours achieved plasma NAP concentrations of 5.2 to 21.0 μg/ml. At these doses, NAPA was less tha impressive in suppressing VPD's at a 90% reduction level in only 2 of 11 patients, with 4 of 11 patients achieving an ≥ 70% reduction. Although efficacy was less than ideal, there were no lupus-like side effects, and the elimination half-life was calculated to be 10.7±2.0 hours, a decided advantage in asymptomatic patients on chronic antiarrhythmic therapy.[15] Similar results were obtained by Roden and his associates who studied 23 patients who had organic heart disease and frequent VPD's. During NAPA therapy, 11 of 23 patients achieved an ≥ 70% reduction in total VPD's as judged by ambulatory electrocardiography. The mean plasma NAPA concentration in those that were responders was 14.3 μg/ml. Patients acetylator phenotype, previous history of drug resistance, and/or the change in QTc interval during therapy were not helpful in predicting which patients might respond to the drug. Of importance was the fact that the elimination half-life was calculated to be 7.5 hours in these patients, and that there were no lupus-like symptoms in any of the 23 patients during therapy. However, 7 of 23 patients had to discontinue medication because of long term toxicity, which was primarily gastrointestinal.[16] Two studies analyzing the electrophysiologic properties of N-acetyl procainamide have concluded that NAPA in human beings is somewhat different from procainamide in that atrial and ventricular refractory periods are increased without increasing cardiac conduction time.[17] Although specific electrophysiologic properties may be different, it appears its antiarrhythmic mechanisms are similar to that of procainamide.[18]

Given the Present State of Knowledge, Should NAPA Be Further Investigated?

Despite theoretical advantages over procainamide, it seems unlikely that NAPA will turn out to be a highly desirable investigational agent since it has a relatively high frequency of undesirable side effects and appears to be less potent than the parent compound.

IV. Quinidine: Is it Still A Reasonable Alternative Choice for the Suppression of Ventricular Ectopy?

Historical Perspective

Quinidine has a special place in the history of ventricular arrhythmias in that it was the first available antiarrhythmic drug recognized effective in suppression of ventricular tachycardia.[19] Although hundreds of authors have discussed its effectiveness in the treatment of ventricular and supraventricular arrhythmias from that time through the 1970's, the quantitative determination of its ability to suppress VPD's awaited studies comparing quinidine to new investigational agents in the late 70's and 80's. Unfortunately, the most recent focus of attention to quinidine has been on its proarrhythmic effects and its ability to initiate torsade de pointes ventricular tachycardia, a topic covered later in this manuscript.

Quantitative Evaluation of Quinidine in Suppressing Ventricular Premature Depolarizations

The most comprehensive evaluation of quinidine was paradoxically in a study designed primarily to look at the efficacy of a new investigational agent, flecainide.[20,21] Baseline arrhythmia frequency in the 139 patients randomized to quinidine therapy was quantitated by two consecutive 24 hour ambulatory ECG's. Dosage adjustment of quinidine up to 1600 mg a day was performed if an 80% suppression of VPD's was not achieved. Based on this flexible dosage schedule, 57% of all patients achieved an $\geq$ 80% VPD reduction. Total suppression of complex forms was achieved in 55% of patients with nonsustained ventricular tachycardia. Fifteen percent of patients discontinued the study before completion because of side effects, primarily gastrointestinal symptoms. Similar results were reported by Sami et al, who analyzed 20 cardiac patients with VPD's in whom quinidine suppressed 80% of VPD's in only 44% of the patients. Total abolition of ventricular ectopy was not achieved in any patient at doses of 300 mg every 6 hours.[22] Side effects, primarily nausea and diarrhea, lead to the reduction in dose in 3 patients. Similar results were obtained by Salerno and his associates in a double blind trial in which quinidine was used for VPD's.[23] The mean VPD reduction for the entire group was 56%, and only 50% of patients had an $\geq$ 80% VPD reduction.

Quinidine Evaluated By Programmed Electrical Stimulation in Patients with Life Threatening Ventricular Tachycardia

The most comprehensive evaluation of quinidine by serial electrophysiologic testing was reported by DiMarco and his associates, who investigated 89 patients with life threatening ventricular tachycardia.[24] During programmed electrical stimulation, 30 of 89 patients had prevention of the initiation of ventricular tachycardia during programmed ventricular stimulation. The mean serum quinidine concentration was 2.9±0.8 μg/ml, essentially identical to the serum level in nonresponders. Nonresponders were characterized by more severe left ventricular dysfunction and use of digitalis.[24] During long term followup, patients initially responding to quinidine remained symptom free and on drug in 84% of the cases. The authors concluded that electrophysiologic testing could select a subset of patients with sustained VT responding to quinidine.

Do Sustained Release Preparations Provide Adequate Plasma Quinidine Concentrations?

The steady state bioavailability of sustained release quinidine has been investigated in multiple studies. In a report of 16 healthy men, the relative bioavailability of quinidine gluconate (Quinaglute®) was compared to that of sustained release quinidine sulfate (Quinidex®).[25] Despite a higher dose of quinidine base in the sulfate preparation, steady state plasma samples at 72 hours of dosing revealed the gluconate preparation provided a slightly higher quinidine plasma level. Both preparations maintained reasonable plasma concentrations at ≥ 8 hours after dose.[25] The ability of the slow release formulation to provide adequate plasma quinidine levels with a half-life allowing q 8 hour dosing has been confirmed by other studies. The importance of quality control in these marketed prolonged release quinidine preparations was emphasized in the report by Myer and his colleagues, who highlighted serious bioavailability problems with the generic prolonged release quinidine gluconate tablet, which was subsequently removed from the market by the U.S. Food and Drug Administration.[26] It appears that the sustained release gluconate and sulfate preparations appear capable of achieving adequate plasma quinidine concentrations on a q 8 hour dosing schedule. In individual patients, however, documentation of efficacy by ambulatory ECG and/or electrophysiologic testing, as well as obtaining adequate plasma quinidine samples are mandatory.

In summary, quinidine appears to be moderately effective in the suppresion of ventricular premature depolarizations in recent studies which have used multiple quantitative ambulatory ECG's. It also appears to be effective in nonsustained ventricular tachycardia and in ventricular tachycardia initiated by programmed electrical stimulation. Its role in sustained ventricular tachycardia or sudden cardiac death is not established.

V. Disopyramide: How Does It Compare to the Other Type IA Agents?

Efficacy in Suppressing Ventricular Premature Depolarizations

An excellent recent review of the electrophysiologic and pharmacokinetic properties of disopyramide can be found in a review by Morady and his colleagues.[27] As with quinidine, the most detailed information with quantitative ambulatory electrocardiography has been obtained in comparison to investigational antiarrhythmic drugs. Our experience with 27 patients with $\geq$ 40 ventricular premature depolarizations per hour on at least two consecutive 24 hour ambulatory electrocardiograms has been reported recently.[28] In this study, disopyramide at a dose of 150 mg p.o. q 6 hours achieved serum trough disopyramide levels averaging 2.2±0.9 µg/ml with peak serum levels 2.6±0.9 µg/ml one hour after steady state dosing. Disopyramide was a rather poor suppressor of ventricular ectopy in this study, achieving a 80% or greater reduction in VPD's in only 22% of patients[28] and an $\geq$ 70% VPD reduction in 36% of patients studied. As with most Class I antiarrhythmics, it was a more potent suppressor of repetetive ventricular forms, resulting in total abolition of nonsustained ventricular tachycardia runs in 12 of 18 patients. Significant side effects occured in one third of patients, frequently necessitating discontinuation of therapy. Using the same dose of disopyramide and a similar study design, Kjekshus and his colleagues studied 25 patients with greater than 1,000 VPD's in a pre-trial 24 hour ambulatory ECG.[29] Their observations parallel ours to a surprising degree. Overall, disopyramide reduced baseline ventricular premature depolarizations only 39%. As with our trial, only 32% of patients receiving disopyramide had an $\geq$ 80% suppression of VPD's. The mean plasma disopyramide concentration in the study was 3.43 µg/ml, slightly higher than in our experience at the same dose range. Although previous clinical trials utilizing minimal ambulatory ECG monitoring appeared more promising, the present trials utilizing more long term quantitative ambulatory ECG information are probably more reflective of the true efficacy of disopyramide in suppressing VPD's.

<u>Is Disopyramide Efficacious in Sustained Ventricular Tachycardia or Ventricular Fibrillation?</u>

A recent study by Lerman and his colleagues reports on 50 patients referred for electrophysiologic testing who had previous documented sustained ventricular tachycardia and/or ventricular fibrillation. Evaluation and assessment of efficacy was made by programmed electrical stimulation.[30] After disopyramide therapy, induction of sustained ventricular tachycardia by programmed electrical stimulation was prevented in 17 of 50 patients (34%). This report appears fairly comparable to similar patient populations in which procainamide and quinidine have been used. Disopyramide doses used in the study ranged from 400-1600 mg per day, resulting in mean plasma disopyramide levels of 3.6±1.2 μg/ml. Plasma level did not predict which patients would respond.

Do Sustained Release Disopyramide Preparations Provide Adequate Bioavailability?

In their study of 12 patients with complex VPD's, Fechter and his colleagues concluded that a long acting (LA) preparation of disopyramide given 300 mg b.i.d. was as effective as the standard disopyramide preparation given 150 mg p.o. q 6 hours.[31] In this study, 75% of the patients had an ≥70% VPD reduction. The mean serum disopyramide concentration of the long acting preparation was 2.8 μg/ml. In a study of disopyramide phosphate (Norpace-CR®) 20 patients were administered 400-800 mg per day of controlled release disopyramide (b.i.d. schedule). Mean serum trough concentrations averaged 2.5±0.9 μg/ml. Forty-five percent of the subjects had adverse side effects, which resulted in discontinuation of the protocol in 25% of the patients.[32] Finally, Nauta and his colleagues have analyzed 33 patients with myocardial infarction to compare plasma levels of disopyramide obtained with conventional capsules and sustained release tablets. This comparison differed in that it was designed to compare a 200 mg t.i.d. regular schedule with a dosage adjustment for those patients whose initial disopyramide plasma concentration exceeded 3.0 μg/ml. Overall, the mean plasma trough concentration for the conventional capsule group was 2.8 μg/ml compared to 3.0 μg/ml in the sustained release group.[33] The ability to suppress ventricular premature depolarizations seemed similar in both groups. In summary, all three studies conclude that long acting disopyramide preparations do provide adequate plasma trough procainamide concentrations given in a b.i.d. dosing schedule and are equally efficacious to the conventional shorter acting oral preparation.

What is the Extent of the Problem of Left Ventricular Dysfunction Precipitated By Disopyramide?

A large volume of literature including multiple case reports has arisen over the past six years regarding the frequency with which disopyramide precipitates congestive heart failure. Based on their own observations and a review of the literature, Podrid and his colleagues reviewed this extensively in 1980.[34] In the high risk population studied by these investigators, the incidence of congestive heart failure during disopyramide was 16%. In patients with pre-existing left ventricular dysfunction, the deterioration can be rapid and serious and result in death.[35] Recent studies using radionuclide ventriculograms demonstrate that disopyramide can suppress global left ventricular function in patients with baseline poor ventricular function (left ventricular ejection fractions less than 35%) at otherwise safe therapeutic serum disopyramide levels.[36,37] This appears to be a much more frequent and serious problem with disopyramide than with procainamide and quinidine. An interesting report recently by Wisenberg and his colleagues help reinforce this observation.[38] Seventeen patients received sequential radionuclide ventriculograms during treatment with all three drugs, separated by adequate washout periods. Of interest was the fact that the rest and exercise left ventricular ejection fraction decreased with all three drugs, with the most prominent deterioration evident with disopyramide.[38] In conclusion, left ventricular dysfunction either measured by noninvasive techniques and/or overtly obvious by precipitation of congestive heart failure is a common and important clinical problem in patients placed in disopyramide therapy, especially those with significant left ventricular dysfunction prior to the initiation of therapy.

VI. Aggravation of Arrhythmia With Type IA Drugs - How Common Is It?

One of the most interesting and clinically important observations in the past decade is the frequency with which antiarrhythmic drugs may become "proarrhythmic". Credit for identifying the potential frequency of the problem belongs to Velebit and his colleagues, who reported on 155 patients who had been referred because of life threatening ventricular arrhythmias, frequently resistant to normal antiarrhythmic therapy.[39] Three criteria were accepted as evidence of drug-induced aggravation of arrhythmia: either a four-fold increase in the hourly frequency of VPD's or a ten-fold increase in the hourly frequency of repetetive ventricular forms, or the development of sustained ventricular tachycardia in a patient in which this was previously not documented. Utilizing these criteria, during 1,024 acute drug testing sequences, the overall frequency of aggravation of arrhythmia was 11%.[39] Importantly, the emergence of sustained ventricular tachycardia and/or ventricular fibrillation in a patient previously not manifesting this serious rhythm disorder was the most common of the three types of arrhythmia aggravation (63% of cases). Relative to this report, quinidine was highest in proarrhythmia frequency (15.4%), with lower frequencies noted for procainamide

(9.1%) and disopyramide (5.9%). Of equal interest was the fact that this frequency of proarrhythmia in the Type IA antiarrhythmic drugs occurs at "therapeutic" mean plasma concentrations.

Can one reasonably extrapolate this frequency of proarrhythmia to the larger majority of patients receiving Type IA antiarrhythmics in a general clinic population rather than a tertiary subspecialty referral center? The answer to that question is clearly no, and the Velebit[39] paper has served to educate clinicians to a potentially lethal problem, but probably has oversensitized them. In fact, the frequency reported in this article is only applicable to similar patients, that is, those with ventricular tachycardia or previous history of sudden cardiac death who have been referred because of drug resistance to the usual antiarrhythmic doses and regimens.

Is There a Subgroup at Highest Risk for "Arrhythmia Aggravation"?

In addition to the important observation by Velebit and his colleagues[39] that quinidine had the highest frequency of arrhythmia aggravation, a large body of literature now exists pointing to the frequency with which quinidine is observed to cause torsade de pointes ventricular tachycardia and ventricular fibrillation.[40] In a review of 31 cases of torsade de pointes ventricular tachycardia due to quinidine, a number of important observations may help us with this question. Importantly, the vast majority of patients in this large case series had quinidine therapy begun for the treatment of atrial flutter and fibrillation. Quinidine doses (mean 1100 mg/day) and quinidine plasma levels (mean 3.7 μg/ml) were in normal therapeutic ranges and therefore not helpful. An important observation for clinical practice is that 74% of the episodes of torsade de pointes ventricular tachycardia occurred within one week of the initiation of therapy. This occurred in a group of patients with the following characteristics: concomitantly on digitalis preparation (90%), electrocardiographic evidence of QT prolongation prior to therapy (71%), and hypokalemia (62%).

To emphasize the population at risk, we reviewed 15 cases of ventricular fibrillation that occurred during 24 hour ambulatory ECG monitoring; the most frequent drug association was the concomitant use of digitalis and quinidine.[41] The mean left ventricular ejection fraction in this group of patients was 34.9%. Similar observations by Denes and his group[42] noted ventricular fibrillation during ambulatory ECG monitoring in 5 patients characterized by severe left ventricular dysfunction, and a prolonged QT interval despite therapeutic plasma levels of quinidine, all of whom were taking digoxin.

In summary, it seems the common theme throughout the variety of reports from many different patient series is the use of quinidine therapy in patients with prolongation of QT interval, characterized by left ventricular dysfunction, with the concomitant use of digoxin. It is important to remember that epidemiologic and clinical studies in the past have documented an association of digitalis use with sudden cardiac death.[41] Therefore, close hospital observation of this group of patients during initiation of quinidine therapy is recommended.

Are any of the Type IA drugs immune from producing torsade de pointes ventricular tachycardia? The answer to that question is no. Multiple case reports document torsade de pointes ventricular tachycardia initiated by procainamide, NAPA, quinidine and disopyramide, in addition to quinidine.[43,44-45,46,47,48,49]

Is There Any Data Suggesting a Reduction in Mortality After Acute Myocardial Infarction Utilizing Type IA Drugs?

This question has been reviewed by Furberg et al.[50,51] All seven randomized, controlled trials utilizing the three clinically available Type IA agents in the early hospital phase of myocardial infarction have failed to demonstrate any effect on survival. Furberg points out multiple problems in study design which primarily relate to small sample size and lack of long term followup.[51] Thus, at the present time there is no evidence that the routine intervention with a Type IA agent in patients with ventricular premature beats after myocardial infarction will result in a reduction in mortality. The author suggests that this uncertainty will continue unless an appropriate study population with adequate sample size and appropriate intervention strategy is developed and tested in a full-scale national trial.[51]

VII. Present Status of Cibenzoline

It is too early in the investigational stage of cibenzoline to make specific conclusions regarding its efficacy relative to other drugs discussed in this section. However, multiple small preliminary studies of 5 to 28 patients indicate potentially comparable efficacy of cibenzoline to procainamide and quinidine in the treatment of VPD's, nonsustained ventricular tachycardia and sustained VT as judged by programmed electrical stimulation.[52,53,54,55,56,57] Furthermore, it appears that its pharmacokinetics may allow a t.i.d. dosing schedule, and it does appear in preliminary trials that plasma cibenzoline concentration is related to the extent of antiarrhythmic efficacy.[57] Cibenzoline merits further investigation because of two other desirable characteristics: it does not appear to have significant negative inotropic effects, and it does not affect the QT or QTc interval, even at toxic levels.[55]

VIII. Preliminary Data for Pirmenol

The paucity of human studies of oral pirmenol looking at antiarrhythmic efficacy are published.[58,59,60] Any realistic comparison with the other Type IA antiarrhythmics in terms of efficacy in treating ventricular arrhythmias is premature. Suffice it to say that the drug appears effective in suppressing VPD's and does so with minimal toxicity and relatively long half-life, which might allow it to be given in a t.i.d. or b.i.d. dosing schedule.[58,59]

REFERENCES

1. Pratt CM, Luck JC, Mann DE, Wyndham CRC: Investigational antiarrhythmic drugs for the treatment of ventricular rhythm disturbances. Cardiol Clin 2:35, 1984
2. Berry K, Garlett E, Bellet S, Gefter W: Use of pronestyl in the treatment of ectopic rhythms. Am J Med 27:431, 1951
3. Karlsson E: Procainamide and phenytoin - comparative study of their antiarrhythmic effects at apparent therapeutic plasma levels. Br Heart J 37:731, 1975
4. Koch-Weser J, Klein S: Procainamide dosage schedules, plasma concentrations, and clinical effects. JAMA 215:1454, 1971
5. Winkle R, Gradman A, Fitzgerald J, Bell P: Antiarrhythmic drug effect assessed from ventricular arrhythmia reduction in ambulatory electrocardiogram and treadmill tests: comparison of propranolol, procainamide and quinidine. Am J Cardiol 42:473, 1978
6. Myerburg R, Kassler K, Keim I, Pefkaros K, Condi C, Cooper D, Castellanos A: Relationship between plasma levels of procainamide, suppression of premature ventricular complexes, and prevention of recurrent ventricular tachycardia. Circulation 64:280, 1981.
7. Buxton A, Roxman H, Marchonski F, Josephson M: Electropharmacology of nonsustained ventricular tachycardia: Effects of Class I antiarrhythmic agents, verapamil and propranolol. Am J Cardiol 53:738, 1984
8. Greenspan A, Horowitz L, Spielman S, Josephson F: Large dose procainamide therapy for ventricular tachyarrhythmia. Am J Cardiol 46:453, 1980
9. Waxman H, Buxton A, Sadowski L, Josephson M: The response to procainamide during electrophysiologic study for sustained ventricular tachyarrhythmias predicts the response to other medications. Circulation 67:30, 1983
10. Myerberg R, Condi C, Sheps D, Appel R, Kiem I, Sung R, Castelanos A: Antiarrhythmic drug therapy in survivors of prehospital cardiac arrest: Comparison of effects of chronic ventricular arrhythmias and recurrent cardiac arrest. Circulation 59:855, 1979
11. Vlasses P, Rocci M, Porini K, Greenspon A, Furguson R: Immediate-release and sustained-release procainamide: Bioavailability at steady state in cardiac patients. Ann Intern Med 98:613, 1983
12. Giardiana E, Fenster P, Bigger JT, Mayersohn M, Perrier D, Marcus F: Efficacy, plasma concentration and adverse effects of a new sustained release procainamide preparation. Am J Cardiol 46:855, 1980
13. Hore P, Bones P, Rolinson T, Ikram H: A pharmacokinetic comparison of two sustained-release oral procainamide preparations. Br J Clin Pharmacol 8:267, 1979
14. Bauer G, Mitchell A, Bates F, Heloestrand K: The assessment of an antiarrhythmic agent, sustained-release procainamide, with the aid of Holter monitoring. Med J Aust 2:733, 1977
15. Winkle R, Jailon P, Kates R, Peters F: Clinical pharmacology and antiarrhythmic efficacy of N-acetylprocainamide. Am J Cardiol 47:123, 1981
16. Roden D, Reele S, Higgins S, Wilkinson G, Smith R, Oates J, Woosley R: Antiarrhythmic efficacy, pharmacokinetics and safety of N-acetylprocainamide in human subjects: Comparison with procainamide. Am J Cardiol 46:463, 1980
17. Jailon P, Rubenson D, Peters F, Mason J, Winkle R: Electrophysiologic effects of N-acetylprocainamide in human beings. Am J Cardiol 47:134, 1981.
18. Sung R, Juma Z, Saksena S: Electrophysiologic properties and antiarrhythmic mechanisms of intravenous N-acetylprocainamide in patients with ventricular disarrhythmias. Am Heart J 105:811, 1983

19. Selzer A: Quinidine in perspective: "The rise and fall of quinidine." Heart Lung 11:20, 1982
20. Hodges M, Salerno D, Granrud G, and the Flecainide-Quinidine Research Group: Flecainide versus quinidine: Results of a multicenter trial. Am J Cardiol 53:66B, 1984
21. Flecainide-Quinidine Research Group: Flecainide versus quinidine for treatment of chronic ventricular arrhythmias in a multi-center clinical trial. Circulation 67:1117, 1983
22. Sami M, Harrison D, Kramer H, Houston N, Shimasaki C, DeBusk R: Antiarrhythmic efficacy of encainide and quinidine: Validation of a model for drug assessment. Am J Cardiol 46:147, 1981
23. Salerno D, Hodges M, Granrud G, Sharkey P: Comparison of flecainide with quinidine for suppression of chronic stable ventricular ectopic depolarizations. Ann Intern Med 98:455, 1983
24. DiMarco J, Garan H, Ruskin J: Quinidine for ventricular arrhythmias: value of electrophysiologic testing. Am J Cardiol 51:90, 1983
25. Taggart W, Holyoak W: Steady-state bioavailability of two sustained-release quinidine preparations: quinidine gluconate versus quinidine sulfate. Clin Ther 5:357, 1983
26. Meyer M, Straughn A, Lieberman P, Jacob J: Serious bioavailability problems with a generic prolonged-release quinidine gluconate product. J Clin Pharmacol 22:131, 1982
27. Morady F, Scheinman M, Desai J: Disopyramide. Ann Intern Med 96:337, 1982
28. Pratt C, Young J, Francis M, Taylor A, Norton H, English L, Mann D, Kopelen H, Quinones M, Roberts R: Comparative effect of disopyramide and ethmozin in suppressing complex ventricular arrhythmias by use of a double-blind, placebo-controlled, longitudinal crossover design. Circulation 69:288, 1984
29. Kjekshus J, Bathen J, Orning O, Storstin L: A double-blind crossover comparison of flecainide acetate and disopyramide phosphate in the treatment of ventricular premature complexes. Am J Cardiol 53:72B, 1984
30. Lerman B, Waxman H, Buxton A, Josephson M: Disopyramide: Evaluation of electrophysiologic effects and clinical efficacy in patients with sustained ventricular tachycardia or ventricular fibrillation. Am J Cardiol 51:759, 1983
31. Fechter P, Ha H, Follath F, Nager F: The antiarrhythmic effects of controlled release disopyramide phosphate and long acting propranolol in patients with ventricular arrhythmias. Eur J Clin Pharmacol 25:729, 1983
32. Zema M: Serum drug concentrations and adverse effects in cardiac patients after administration of a new controlled-release disopyramide preparation. Ther Drug Monit 6:192, 1984
33. Nauta I, Van de Calseyde J, Hertzberger D: Plasma levels of disopyramide after administration of conventional capsules of sustained-release tablets. Curr Med Res Opin 8:582, 1983
34. Podrid P, Schoeneberger A, Lown B: Congestive heart failure caused by oral disopyramide. N Eng J Med 302:614, 1980
35. Desai J, Scheinman M, Hirschfeld D, Zelis R, Peters R: Cardiovascular collapse associated with disopyramide therapy. Chest 79:545, 1981
36. Kowey P, Friedman P, Podrid P, Zielong J, Lown B, Woinn J, Hallman B: Use of radionuclide ventriculography for assessment of changes in myocardial performance using disopyramide phosphate.Am Heart J 104:769, 1982
37. Greene A, Iskandrin A, Hokki A, Kane S, Segeal B: Effective role of disopyramide therapy on left ventricular function. Chest 83:480, 1983

38. Wisenberg G, Zawadowski G, Gebhardt B, Prato F, Goddard M, Nichol P, Rechnitzer P, Gryfe-Becker B: Effects of ventricular function of disopyramide, procainamide and quinidine as determined by radionuclide angiography. Am J Cardio 53:1292, 1984
39. Delibat V, Podrid P, Laun B, Cohen B, Grabois T: Aggravation and provocation of ventricular arrhythmias by antiarrhythmic drugs. Circulation 65:886, 1982
40. Bauman J, Bauernfeind R, Hoff J, Strasberg B, Swiryn S, Rosen K: Torsade de pointes due to quinidine: Observations in 31 patients. Am Heart J 107:425, 1984
41. Pratt C, Francis M, Luck J, Wyndham C, Miller R, Quinones M: Analysis of ambulatory electrocardiograms in 15 patients during spontaneous ventricular fibrillation with special reference to preceding arrhythmic events. JACC 2:789, 1983.
42. Denes P, Gabster A, Huang S: Clinical, electrocardiographic and follow-up observations in patients having ventricular fibrillation during Holter monitoring. Am J Cardiol 48:9, 1981
43. Chow M, Piergies A, Bowsher D, Ruffi J, Krichner W, Ruo T, Assada A, Kawana J, Adkinson A: Torsade de pointes induced by N-acetylprocainamide. JACC 4:621, 1984
44. Olshanski B, Martins J, Clint S: N-acetylprocainamide causing torsade de pointes. Am J Cardiol 50:1439, 1982
45. Commerford P, Beck W: Ventricular tachycardia with torsade de pointes morphology induced by oral disopyramide. S Afr Med J 58:447, 1980
46. Chia B: Disopyramide induced atypical ventricular tachycardia. Aust NZ J Med 10:665, 1980
47. Lowe K, Gantz K, Stetson P, Lucchesi B, Pitt B: Disopyramide-induced ventricular tachycardia. Arch Intern Med 140:413, 1980
48. Croft C, Kennelly B: Ventricular tachyarrhythmias induced by disopyramide and other similar anti-arrhythmic drugs. S Afr Med J 59:871, 1981
49. Wald R, Waxman M, Colman J: Torsade de pointes ventricular tachycardia: A complication of disopyramide shared with quinidine. J Electrocardiol 14:301, 1981
50. May G, Ederlein K, Furberg C, Passamani E, DeMets D: Secondary prevention after myocardial infarction: A review of long-term trials. Prog Cardiovasc Dis 4:331, 1982
51. Furberg C: Effective antiarrhythmic drugs on mortality after myocardial infarction. Am J Cardiol 52:32C, 1983
52. Browne K, Prytowski E, Zipes D, Chilson D, Heger J: Clinical efficacy and electrophysiologic effects of cibenzoline therapy in patients with ventricular arrhythmias. JACC 3:857, 1984
53. Kostis J, Krieger S, Morayra A, Cosgrove N: Cibenzoline for treatment of ventricular arrhythmias: A double-blind placebo-controlled study. JACC 4:372, 1984.
54. Cocco G, Strozzi C, Pansini R, Rochat N, Boulgarelli R, Padula A, Sfrasi C, Yassini A: Antiarrhythmic use of cibenzoline, a new Class I antiarrhythmic agent with Class III and IV properties in patients with recurrent ventricular tachycardia. Eur Heart J 5:108, 1984
55. Kushner M, Magirose, Peters R, Carliner N, Plotnick G, Fisher M: The electrophysiologic effects of oral cibenzoline. J Electrocard 17:15, 1984
56. Canal M, Flouvat B, Trenblay D, DuFour A: Pharmacokinetics in man of a new antiarrhythmic drug, cibenzoline. Eur J Pharmacol 24:509, 1983
57. Brazzell R, Aogaichi K, Heger J, Somberg J, Carliner N, Morganroth J: Cibenzoline plasma concentration and antiarrhythmic effect. Clin Pharmacol Ther 35:307, 1984

58. Hammil S, Shand D, Rutledge P, Hindman M, Baker J, Pritchett E: Pirmenol, a new antiarrhythmic agent: Initial study of efficacy, safety and pharmacokinetics. Circulation 65:369, 1982
59. Anderson K, Lutz J, Nappi J: Pirmenol for control of ventricular arrhythmias: oral dose-ranging and short-term maintenance study. Am J Cardiol 53:522, 1984
60. Anderson J, Lutz J, Sanders S, Nappi J: Efficacy of intravenous pirmenol hydrochloride for treatment of ventricular arrhythmias: A controlled comparison with lidocaine. J Cardiovasc Pharmacol 5:213, 1983.

6

Mexiletine, Tocainide and Ethmozine: Newer Class I Antiarryhthmic Agents

Allan M. Greenspan, M.D.

INTRODUCTION

Class I antiarrhythmic agents are characterized by suppression of the maximum rate of rise ($\dot{V}_{max}$) and amplitude of the cardiac action potential which results in impaired conduction velocity in myocardial tissue.[1] Overall clinical antiarrhythmic efficacy in this group is dependent upon the ability of these agents to alter conduction and refractoriness, and suppress diastolic depolarization, thereby interfering with reentrant circuits and suppressing automatic foci.

Sub-classification of this group of agents is based upon the intensity of suppression of the $\dot{V}_{max}$ and the effect on action potential duration[2] (See Table 1). IA agents have an intense suppressant effect on V_{max} and prolong the action potential, while IB agents have a milder suppressant effect on $\dot{V}_{max}$ and shorten the action potential. The IB agents, although not as potent suppressors of conduction as the IA agents, don't have the proarrhythmic properties of the IA agents which may be related to their action potential lengthing effects and QT prolongation. Thus, the development of a group of unclassified Class I agents

Table 1

Class I Antiarrhythmic Agents

Subgroup	V_{max}	AP duration
A	↓↓	↑
B	↓	↓
C	↓↓↓	→
UC	↓↓	↓

UC = unclassified

which combine properties of both the IA and IB agents, with the intense suppressant effect on $\dot{V}_{max}$ of the IA group and the action potential shortening effect of the IB group, has generated a great deal of interest and enthusiasm.

This paper will focus on the pharmacokinetics, antiarrhythmic properties, and side effects, of three of the newer Class I antiarrhythmic agents, mexiletine and tocainide of the IB group and ethmozin of the hybrid unclassified group, and will compare and contrast the characterisitcs of the IB and unclassified hybrid group.

Mexiletine

Mexiletine is a phenoxy substituted primary amine whose chemical structure and membrane stabilizing effect are quite similar to lidocaine[3] (Figure 3). Its pharmacokinetics differ significantly from that of lidocaine, allowing a high degree of GI absorption and thus achieving significant plasma levels with oral administration.

Figure 1

MEXILETINE

CHEMICAL STRUCTURE

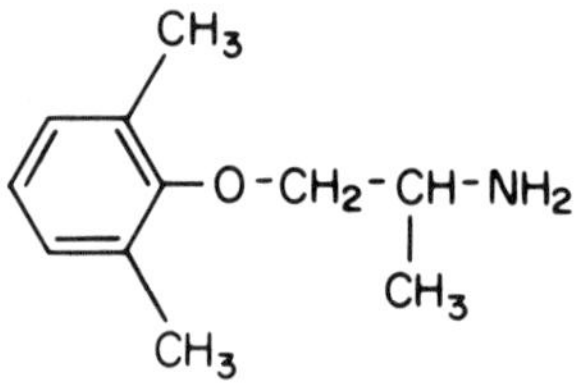

In normal individuals, following an oral dose, almost 100% of the drug is absorbed, achieving peak plasma levels in 2-4 hours.[4] There is delayed and sometimes incomplete absorption of mexiletine in patients following acute myocardial infarction. The mean steady state plasma concentrations of mexiletine that are achieved vary directly with dose between 600 to 1000mg daily.[5] Oral regimens of these dosages were able to achieve plasma concentrations in the therapeutic range of 1-2 mcg/ml.[6]

Following IV dosages, there is a rapid fall off of plasma concentrations compatible with a tissue uptake or redistribution phase.[4] Pharmacokinetic studies indicate that the volume of distribution for the drug is relatively large (7-10 l/kg) and the elimination half time varies with the clinical condition of the patient, being approximately 8-10 hours in normals but may be increased to 12-16 hours in patients post-myocardial infarction.[5]

The major mode of elimination is hepatic metabolism followed by urinary excretion of the metabolites with only 8% of the drug being excreted unchanged in the urine. Urinary pH strongly affects elimination of mexiletine, Since its metabolites are alkaline compounds that are actively secreted by the renal tubule. The pKa of the metabolites is in the range of 4-8. Thus acidification of the urine causes ionization of the metabolites, decreasing their reabsorption from the tubule and therefore increasing excretion signfificantly.[7]

As with pharmacokinetics, hemodynamic effects of the drug are dependent on the clinical condition of the patient. In normal individuals there is no perceptible negative inotropic effect of mexiletine. In patients with coronary artery disease and angina, intravenous mexiletine caused mild increases in heart rate (4-5 beats/minute) as well as mild increases in pulmonary artery systolic pressure (3-4mmHg), systolic arterial pressure (4 mmHg) and left ventricular end diastolic pressure (3-5 mmHg). There was a mild reduction in cardiac output (0.5 l/min) and in dp/dt with isometric exercise in coronary artery disease patients following mexiletine. The fall in cardiac output in coronary disease patients while on mexiletine has been attributed to both an increase in afterload and a direct suppression of dp/dt.[8]

Cellular electrophysiologic effects of mexiletine are that of a classic Group IB agent. Mexiletine has been found to reduce $\dot{V}_{max}$ of the cardiac Purkinje fiber action potential as well as phase 4 diastolic depolarization and the action potential duration. In addition mexiletine suppresses membrane responsiveness. Mexiletine also increases the ventricular fibrillation threshold.[9]

The clinical electrophysiologic effects of mexiletine like its hemodynamic effects are dependent once again on the patients underlying organic heart disease. In normal individuals mexiletine has no significant effect on any electrophysiologic parameters of sinus node. AV node or infranodal conduction or atrial and ventricular refractoriness.[10] However in patients with coronary artery disease and evidence of abnormal impulse formation or conduction disturbances, mexiletine does alter parameters of sinus node and AV conduction system function. In patients with underlying sinus node dysfunction mexiletine has been shown to increase the sinus node recovery time and in patients with abnormal AV nodal function, mexiletine has increased AV nodal conduction time, the cycle length to Wenckebach, and AV nodal function refractory period. Also in patients with underlying organic heart disease mexiletine has been shown to increase the HV interval by as much as 11 msec and increase the His-Purkinje system relative refractory period.[11]

Like other IB agents and, to some extent the IA agents mexiletine can have substantial clinical efficacy despite minimal effects on clinical electrophysiologic parameters. Furthermore mexiletine seems to be particularly effective in patients with underlying transient metabolic or ischemic derangements.

In the setting of actue ischemia, a study by Achuff[12] indicated that oral mexiletine was effective in suppressing complex ventricular ectopy and ventricular tachycardia or fibrillation. Ninety-seven patients presenting with acute MIs were randomized to either oral mexiletine or placebo. Mexiletine reduced the incidence of ventricular tachycardia and fibrillation and complex ectopy (R on T phenomena) by 44% compared with placebo.

In the setting of stable organic heart disease however the efficacy of mexiletine is somewhat more variable depending upon the arrhythmia being treated.

Using a standard for drug efficacy of >70% reduction in mean VPD frequency, mexiletine effectively suppressed isolated VPDs in anywhere from 15-83% of patients in a series of studies based on Holter evaluation (see Table 2).[5],[6],[13],[14],[15] Suppression of complex ventricular ectopic activity (bigeminy, couplets) defined as either >95% or 100% reduction in couplets was achieved in 26-50% of patients.[3],[14] For non-sustained ventricular tachycardia, judged by Holter evaluation, mexiletine abolished the arrhythmia in anywhere from 0%-80% of patients.[3],[5],[6],[14],[16]

Table 2

Mexiletine

Clinical Efficacy (% Suppression)*

Author	VPDs	Complex VEA	VT
Mehta et al. (AJC 1980)	10/12(83%)	6/12(50%)	4/ 5(80%)
Heger et al. (AJC 1980)	2/13(15%)	--------	0/ 9(0%)
Singh et al. (AJC 1984)	18/26(69%)	--------	--------
Waspe et al. (AJC 1983)	4/12(33%)	--------	3/33(9%)
Plileo et al. (AJC 1982)	----------	--------	1/11(9%)
Greenspan et al. (The Pharm. 1983)	----------	--------	0/11(0%)
Jewitt et al. (Postgrad Med. J. 1977)	13/25(52%)	5/19(26%)	6/14(43%)

* VPD suppression - >70% reduction
complex VEA suppression - >90% reduction
VT suppression - 100% reduction

Using programmed ventricular stimulation in with recurrent sustained ventricular tachycardia and stable organic heart disease mexiletine was found to be somewhat less effective, although most antiarrhythmic agents show lower efficacy using this more stringent mode of evaluation. For patients with inducible sustained ventricular tachycardia, the efficacy rate of mexiletine varied from 0% to 19%.(15),(16),(17),(18) If polymorphic ventricular tachycardia or ventricular fibrillation were included along with sustained ventricular tachycardia the incidence of efficacy increased substantially to 23%.(15)

Comparative studies of the efficacy of mexiletine versus standard Group 1A agents in suppressing complex ventricular ectopy and non-sustained ventricular tachycardia help place these efficacy rates for mexiletine in some perspective. Jewitt(14) compared the efficacy of oral mexiletine and oral procainamide against isolated VPDs, complex forms and non-sustained ventricular tachycardia. There were no significant differences in the efficacy rates for both of these agents in suppressing isolated VPDs by >75% of control (mexiletine 52%, procainamide 60%), in the erradication of couplets (mexiletine 26%, procainamide 37%), and in the abolition of non-sustained ventricular tachycardia (mexiletine 43%, procainamide 60%). Similarly Singh et al.[13] compared the efficacy of oral mexiletine and oral quinidine by Holter evaluation in patients with frequent, complex ventricular ectopic activity, and non-sustained ventricular tachycardia. They also found no significant differences in the efficacy rate of these two agents in reducing isolated VPD frequency by >70% (mexiletine 70%, quinidine 69%) or an eradication of non-sustained ventricular tachycardia (mexiletine 72%, quinidine 71%).

One recent observation indicates that in patients whose ventricular tachyarrhythmias are unresponsive to mexiletine alone, the combination of mexiletine and other antiarrhythmic agents, particularly Group IA agents may markedly increase antiarrhythmic efficacy. Duff, et al[19] found that the combination of mexiletine and quinidine suppressed non-sustained as well as sustained ventricular tachycard and ventricular fibrillation by Holter evaluation in 94% of patients whereas quinidine alone was effective in only 35% of patients and mexiletine alone in 69% of patients. Using the more stringent criteria of programmed stimulation Waspc et. al.[16] found that combining mexiletine plus additional antiarrhythmic agents prevented induction of sustained VT in 15% of patients whereas mexiletine alone was effective in only 9% of patients. Finally in a group of 24 patients with recurrent sustained ventricular tachycardia or fibrillation inducible by programmed stimulation who failed therapy with mexiletine alone and quinidine or procainamide alone, Greenspan et. al.[20] found the combination of mexiletine and a 1A agent prevented induction of sustained ventricular tachyarrhythmia in 38% of patients.

Side effects with mexiletine are not inconsiderable (see Table 3). Like the other 1B agents the major side effects involve the central nervous system (tremor, dizziness, gait abnormalities, blurred vision) and the gastrointestinal tract (nausea, vomiting, dyspepsia, gastritis). Overall incidence of side effects varied from 47% to 81% but the reported incidence of intolerable side effects requiring change in drug therapy was substantially lower, 8-31%.[13][14] The incidence of GI side effects was approximately 39% and that for CNS side effects was 35-50%.[13][16] The incidence of side effects appears to be plasma level dependent as the overall incidence is significantly higher at plasma levels at >1.5-2.0mg/ml.[21]

Table 3. Mexiletine Side Effects

Sx/Signs	Frequency	
(1) CNS	45%	
dizziness		30%
tremor		29%
ataxia		17%
(2) GI	44%	
dyspepsia		50%
nausea		32%
vomiting		14%
(3) Cardiovascular		2-9%
bradycardia		
hypotension		
(4) Side effects requiring drug withdrawal		
CNS		8%
GI		8%

Tocainide

Tocainide, like the other Group 1B antiarrhythmic agents, has a substituted primary amine structure (see Figure 2). It differs from that of lidocaine by a de-ethylation which results in loss of the high first pass liver extraction, and thus allows achievement of effective plasma levels following oral administration. Pharmacokinetics of tocainide are not too dissimilar from those of mexiletine.

Figure 2

TOCAINIDE HCL: CHEMICAL STRUCTURE

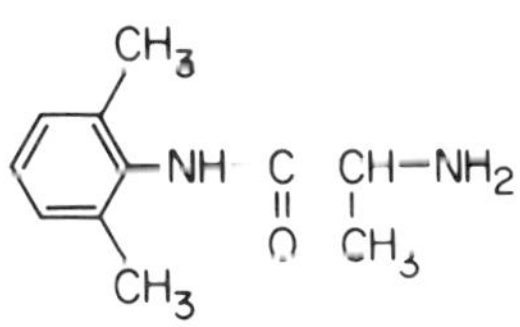

TOCAINIDE

Following an oral dose, peak plasma levels of tocainide are achieved in 1-2 hours. The volume of distribution is relatively large (2.4-3.8 l/kg) and the drug is 50% protein bound.[22] The major routes of elimination are by urinary excretion (40% of the oral dose) and metabolic hepatic degradation to the glucuronide (60% of the administered dose).[23] Consequently the half time for elimination is greatly dependent on intact renal function being 12-14 hours in patients with normal renal function[23],[24] but increased to 27 hours in patients with impaired renal function.[24]

As with mexiletine the hemodynamic effects of this 1B agent are dependent on the degree of underlying organic heart disease. In patients with normal cardiac function there are no significant hemodynamic effects up to a mean plasma concentration of 4.6 mg/l.[25] In patients with chronic stable coronary artery disease, tocainide causes mild reduction in left ventricular systolic function but primarily at high plasma levels (11-18 mg/l).[26] In patients with compensated left ventricular failure, tocainide causes small increases in left ventricular end diastolic pressure but no discernable decrease in overall systolic function. In patients immediately post acute myocardial infarction, low doses of tocainide have caused significant increases in systemic vascular resistance, mean systolic blood pressure and have resulted in concomitant decrease in cardiac output and stroke volume due to increased afterload.(27)

The cellular electrophysiologic effects of tocainide are quite similar to that of mexiletine. Tocainide shows mild to moderate suppression of $\dot{V}_{max}$, of sheep Purkinje fiber action potentials, which is greater than that of lidocaine. Also it shows a mild suppression of the action potential amplitude but produces significant reduction in phase 4 diastolic depolarization thereby suppressing automaticity. Tocainide shortens the action potential duration and also shortens the refractory period of ventricular muscle but to a lesser extent, leading to an increase in the ratio of the effective refractory period to action potential duration. Like mexiletine tocainide also suppresses membrane responsiveness, particularly at concentrations greater than 10mg/l and increases the ventricular fibrillation threshold.(21),(28)

The clinical electrophysiologic effects of tocainide are dependent on the presence of underlying organic heart disease. In patients with normal cardiac function tocainide has no significant effects on sinus node or AV nodal function, on intra-atrial and His-Purkinje conduction, and does not significantly alter atrial and ventricular effective refractory periods. However in patients with underlying organic disease and left ventricular dysfunction, tocainide decreases the atrial effective refratory period but has inconsistent effects on AV nodal effective refractory period. Like mexiletine and lidocaine, tocainide markedly prolongs antegrade effective refractory period of accessory pathways and is therefore quite useful in treating arrhythmias associated with WPW.

The clinical efficacy of tocainide like the other IB agents is limited primarily to ventricular ectopic activity, and ventricular tachyarrhythmias.(See Table 4) As in the case of mexiletine, tocainide has particular efficacy in these types of arrhythmias that occur in the setting of transient metabolic or ischemic derangements. In fact tocainide was found to be equally effective as lidocaine in suppressing complex ventricular ectopic activity, ventricular tachycardia and ventricular fibrillation in the acute MI setting.[30] Similarly, in a recent study of patients having complex ectopic activity and ventricular tachycardia post-cardiac surgery, tocainide was found to be more effective than lidocaine in suppressing

Table 4

Tocainide

Clinical Efficacy (% Suppression)*

Author	VPDs	VT
Winkle et al. (Circ. 1976)	11/15(73%)	--------
Woosley et al. (Circ. 1977)	8/12(67%)	--------
LeWinter et al. (AJC 1980)	5/10(50%)	--------
Emergency Use Protocol	680/1236(55%)	65/77(84%)
Winkle et al. (Circ. 1978)	--------	8/13(62%)
Hatagee et al. (Clin. Cardiol. 1983)	12/20(60%)	7/20(35%)
Ryan et al. (AJC 1978)	--------	14/21(67%)

* VPD suppression - >70% reduction
VT suppression - 100% reduction

complex ventricular ectopic activity (tocainide 75%, lidocaine 60%) and non-sustained ventricular tachycardia (lidocaine 75%, tocainide 90%).[31]

In patients with chronic stable organic heart disease and ventricular ectopic activity tocainide shows reasonably high efficacy rates for suppression of isolated VPDs both in single blind placebo controlled studies and in comparative studies with 1A agents. Using greater than 75% reduction in mean VPD frequency for isolated VPDs the efficacy rates for tocainide varied between 35 to 73% in a series of single blind placebo controlled studies examining drug efficacy.[32],[33],[34] When compared with procainamide, tocainide was almost equally effective in suppressing isolated VPDs by > 70% (tocainide 60%, procainamide 80%).[35] In another comparative study between tocainide and quinidine for suppression of chronic high frequency VPDs and complex ventricular ectopic activty, the efficacy (defined as >75% reduction in mean VPD frequency and erradication of complex forms) of tocainide and quinidine were not significantly different[36] (tocainide 37% and quinidine 50%).

For therapy of recurrent sustained ventricular tachyarrhythmias evaluated by passive recording techniques and clinical evaluation, Winkle et. al.[33] noted a 60% success rate in suppression of symptomatic recurrences or erradication of ventricular tachycardia by Holter[37] and Ryan et. al.[38] noted a 67% erradication of recurrent sustained ventricular tachycardia by Holter[38]. Finally, pooled data from a number of different groups in the emergency use protocol indicates that in 77 patients with recurrent ventricular tachycardia or ventricular fibrillation or and syncope or pre-syncope, tocainide prevented recurrences by Holter or symptoms in some 84% of patients.[39]

Like the other 1B agents side effects with tocainide are relatively frequent and involve the central nervous system (dizziness, paraesthesia, tremor) and GI systems (nausea and vomiting) (see Table 5). In addition tocainide shows cardiovascular side effects more commonly associated with the unclassified Group 1 or 1C agents i.e. bradycardia and hypotension particularly with IV infusion. With IV dosing of tocainide side effects occur in approximately 16% of patients; GI side effects in 12.5% and bradycardia and hypotension in 7%.[39] Much less common but more serious side effects which are somewhat idiosyncratic in nature include restrictive pulmonary disease, drug fever, and most importantly aplastic anemia and agranulocytosis.

Table 5. Tocainide Side Effects

	Sx/Signs	Frequency	Required Withdrawal
(1)	CNS	35%	6%
(2)	GI	35%	3%
(3)	Rash	20%	2%
(4)	Cardiovascular	7%	3%
(5)	Idiosyncratic	<1%	100%
	Pulmonary Fibrosis		
	Fever		
	Marrow Suppression		
	Pro-arrhythmic		

Like mexiletine side effects of tocainide tend to be dose related with an apparent threshold phenomena.(21)(39) At plasma tocainide levels of >8mg/l all adverse effects have a 55% incidence, while below 8mg/l adverse effects have only a 35% incidence. One other serious side effect which is more commonly seen with the unclassified group 1 agents, particularly aprindine, is a pro-arrhythmic effect manifested by increased frequency of ventricular fibrillation in acutely ischemic myocardium under conditions of more rapid heart rates.(37)

Ethmozin

Ethmozin belongs to the unclassified Group 1 agents whose electrophysiologic effects are a hybrid of the 1A and 1B agents.(21) Its chemical structure is also quite different from that of the 1B agents which are phenal substituted amine compounds. Ethmozin instead is a phenothiazine derivative with a tricyclic structure making it chemically more akin to the tricyclic antidepressants (see figure 3).(21)

Figure 3

ETHMOZIN

CHEMICAL STRUCTURE

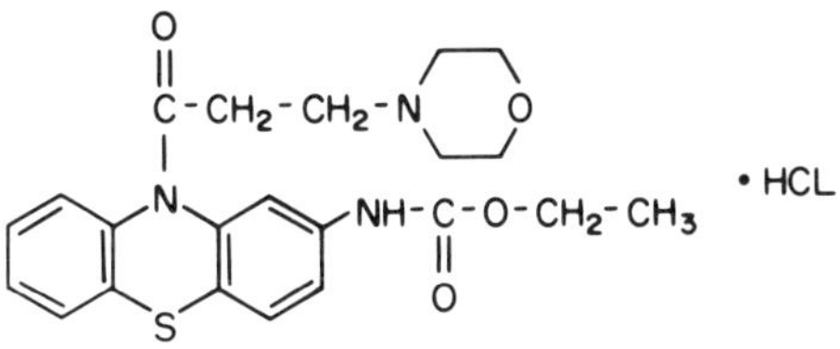

Ethmozin is rapidly absorbed from the gastrointestinal tract with peak plasma levels achieved at 45-90 minutes following an oral dose . As ethmozin is 92% bound to plasma proteins, it is not surprising that any excretion of the intact compound does not play a major role in its elimination. Rather ethmozin is almost completely metabolized by the liver and less than 1% is excreted in the urine and feces as the intact drug. The elimination half time for ethmozin is quite variable, depending upon the underlying condition of the patient; being 2-5 hours in normal individuals, up to 17 hours in patients with organic heart disease, to 48 hours in patients with renal insufficiency. The drug also has a relatively large volume of distribution of 12-20 l/kg which is independent of the patient's underlying clinical condition.(40),(41)

Hemodynamic effects of ethmozin have been extensively tested only in animals. To date ethmozin has shown no significant effects on autonomic tone, left ventricular ejection fraction or dp/dt in dogs. In one study performed in the cardiac catheterization laboratory in humans, ethmozin showed no effects on intra-cardiac pressures, but did show slight to moderate reductions in cardiac output in 4 of the 6 patients.(41)

Cellular electrophysiologic effects of ethmozin are not dissimilar from the 1B agents. Ethmozin shows a dose depedent suppression of the $\dot{V}_{max}$ in Purkinje fibers, and at high doses ethmozin reduces maximum diastolic potential and action potential amplitude. Although ethmozin has been shown to have no effect on the slow phase 4 diastolic depolarization in normal Purkinje fibers, ethmozin does block abnormal automaticity induced by epinepherine, barium or carnitine palmitate. Like the Group 1B antiarrhythmic agents ethmozin does accelerate phase 2 and 3 of repolarization leading to a shortening of the action potential. It also decreases the effective refractory period of Purkinje fibers and decreases membrane responsiveness.(41),(42) Interestingly the suppressant effect of ethmozin on V_{max} in Purkinje fibers differs mechanistically from that of procainamide and quinidine. Rather than altering reactivation of the fast Na current as do the IA agents, ethmozin causes a direct reduction of the maximal conductance of the sodium channel.(21) Finally, ethmozin prolongs conduction velocity in ventricular muscle under ischemic conditions following LAD occulusion in dogs.(21) This has been shown to increase the incidence of ischemically induced ventricular fibrillation just as in the case of aprindine and tocainide.

There have been few studies on the clinical electrophsyiologic effects of ethmozin in man. These studies indicate that ethmozin has no effect or mildly increases the basic sinus cycle length the PR and QRS intervals and has no effect on the QT interval. Two studies have examined the effects of ethmozin on electrophysiologic parameters in man. One study noted no effect on sinus node function, AV nodal or His-Purkinje conduction characteristics or atrial and ventricular refractory periods.(43) While the second study observed prolongation of most of these parameters following ethmozin.(44)

Unlike the group IB agents, ethmozin shows significant antiarrhythmic efficacy against supraventriuclar premature depolarizations (SVPDs) as well as chronic ventricular premature depolarizations, complex forms and non-sustained ventricular tachycardia. In a study of 14 patients, ethmozin, at doses of 1.5 to 15 mg/Kg/day, reduced mean SVPD frequency by >65% in 80% of the patients studied.(45)

The efficacy of ethmozin against chronic isolated VPDs in patients with stable organic heart disease is moderately high (see table 6). Various reports

Table 6

Ethmozin

Clinical Efficacy (% suppression)*

Author	VPDs	Complex VEA	VT
Kennedy (Circ. 1978)	7/12(58%)*	-----------	--------
Podrid (Circ. 1980)	14/26(54%)*	5/ 6(83%)·	2/2(100%)+
Podrid (AJC 1982)	--------	---------	32/58(56%)+
Pratt (AJH 1983)	28/39(72%)**	26/31(84%)·	15/19(79%)+
Pratt (Circ. 1984)	17/27(63%)**	17/27(63%)·	11/16(69%)+

* suppression = >63% reduction
** suppression = >70% reduction
· suppression = > 90% reduction
+ suppression = 100% reduction

indicate an incidence of effective VPD suppression (defined as >63-70% reduction in mean frequency) ranging from 54-72%.[46],[47],[48],[49] For complex ventricular ectopy the incidence of >90% suppression varied between 63-87%[47],[48],[49], while for abolition of non-sustained ventricular tachycardia the incidence varied between 56-79%[47],[49],[50]. To date, only one study has examined the efficacy of ethmozin in inducible sustained ventricular tachycardia using electrophysiologic drug testing. In this small group of patients ethmozin failed to prevent induction of the ventricular tachycardia in all cases, though it did noticably prolong the tachycardia cycle length.[44]

Two studies have compared the relative efficacy of ethmozin with standard group 1A agents. Morganroth et al.[51] showed that in patients with stable organic heart disease and chronic SVPDs and VPDs that ethmozin and quinidine were of equivalent efficacy in suppressing SVPDs (ethmozin 87%, quinidine 84%) and VPDs (ethmozin 80%, quinidine 83%), where efficacy was defined as >70% reduction in mean frequency. Pratt et al.[49] demonstrated that ethmozin showed a higher efficacy than disopyramide in suppressing isolated VPDs by >70% (ethmozin 63%, disopyramide 41%) and ventricular couplets by >90% (ethmozin in 63%, disopyramide 44%) but was equally efficacious as disopyramide in erradication of non-sustained ventricular tachycardia (ethmozin 69%, disopyramide 67%).

Side effects with ethmozin are minimal (see Table 7), and fall into 2 major categories: central nervous system and gastrointestinal.[21] Among CNS side effects, lightheadedness (12%) and headache (7%) are most frequent; weakness, nervousness and tremor occurring in <3% of cases. GI side effects are slightly less frequent. Nausea (7%) is the most common GI side effect, with diarrhea dyspepsia and constipation occurring in < 2.5% of cases. In general the side effects are well tolerated, and rarely require alteration in therapy.

Summary

To summarize, the class 1B agents as represented by mexiletine and tocainide show mild suppressant effects on V_{max} and have no action potential or QT prolongation effect thus making them relatively safe antiarrhythmic agents to use (no pro-arrhythmic effect). They have very high efficacy for ventricular ectopic activity and non-sustained ventricular tachycardia in the setting of acute ischemia or metabolic derangements, but have only mild to moderate

Table 7. Ethmozin Side Effects

	Sx/Signs	Frequency
(1) CNS		
	lightheadedness	12%
	headache	7%
	weakness	3%
	nervousness	2%
(2) GI		
	nausea	7%
	diarrhea	3%
	dyspepsia	2%
	constipation	1%

efficacy against these arrhythmias with chronic stable organic heart disease. On the other hand ethmozin exemplifying one of the unclassified group 1 agents has cellular electrophysiologic effects that are hybrid of the 1A and 1B agents showing more intense suppression of V_{max} but again no prolongation of the action potential or QT. Thus to date ethmozin has not shown evidence of significant pro-arrhythmic effect such as the 1A agents but like the 1A agents has a higher efficacy than the Group 1B agents against ventricular ectopic activity and ventricular tachyarrhythmias in the setting of stable organic heart disease. Thus it is likely that the unclassified Group 1A agents may hold a great deal of promise for the future and may take us one step closer to the "ideal" antiarrhythmic agent which has a 100% efficacy and 0% toxicity.

REFERENCES

1. Vaughan Williams EM: Classification of antiarrhythmic drugs. Pharm. Ther. 1:115, 1975.
2. Vaughan Williams EM: Antiarrhythmic Action. Academic Press, London 1980.
3. Mehta J and Conti RC: Mexiletine, a new antiarrhythmic agent for treatment of premature ventricular complexes. Am. J. Cardiol: 49: 455-60, 1982.
4. Prescott LF, Pottage A, Clements JA: Absorption, distribution and elimination of mexiletine. Postgrad. Med. J. 53: (Suppl 1): 50-55, 1977.
5. Heger JJ, Nattel S, Rinkenberger RL, Zipes DP: Mexiletine therapy in 15 patients with drug resistant ventricular tachycardia. Am. J. Cardiol. 45:627-32, 1980.
6. Talbot RG, Julian DG, Prescott LF: Long-term treatment of ventricular arrhythmias with oral mexiletine. Am. Heart J. 91: 58-65, 1976.
7. Kay CM, Kiddie MA, Turner P: Variable pharmacokinetics of mexiletine. Postgrad. Med. J. 53:(Suppl 1): 56-62, 1977.
8. Banim SO, DaSilva A, Stone D et. al. Observations of the hemodynamics of mexiletine. Postgrad Med. J. 53:(Suppl): 74-6, 1977.
9. Danneberg PB, & Shelley JH: The pharmacology of mexiletine. Postgrad. Med. J. 53(Supp.1): 25-32, 1977.
10. Roos JC, Paalman ACA, Dunning AJ: Electrophysiologic effects of mexiletine in man. Br. Heart J. 38:62-72, 1976.
11. McComish M, Robinson C, Kitson D et al.: Clinical electrophysiologic effects of mexiletine. Postgrad. Med. J. 53(Suppl. 1) 85-91, 1977.
12. Achuff SC, Pottage A, Prescott L, Campbell RWF, Murray A, Julian DG: Mexiletine in the prevention of ventricular arrhythmias in acute myocardial infarction. Postgrad. Med. J. 53: 163-4, 1977.
13. Singh JB, Rasul AM, Shah A, Adams E, Flessas A, Kocot SL: Efficacy of mexiletine in chronic ventricular arrhythmias compared with quinidine: a single-blind, randomized trial. Am. J. Cardiol. 53: 84-87, 1984.
14. Jewitt DE, Jackson G, McCormish M: Comparataive anti-arrhythmic efficacy of mexiletine, procainamide and tolamolol in patients with symptomatic ventricular arrhythmias. Postgrad. Med. J. 53(suppl 1): 150-3. 1977.

15.Greenspan AM, Lowenthal DT, Spielman SR, Horowitz LN: Efficacy of Mexiletine in ventricular tachyarrhythmics. The Pharmacologist 25:(3) 246, 1983.
16. Waspe LE, Waxman HL, Buxton AE, Josephson ME: Mexiletine for control of drug-resistant ventricular tachycardia: clinical and electrophysiologic results in 44 patients. Am. J. Cardiol. 51: 1175-81, 1983.
17. Palileo E, Welch W, Hoff J, Strasberg B, Bauernfeind R, Swiryn S, Coelho A, Rosen K: Lack of effectiveness of oral mexiletine in patients with drug refractory, paroxysmal sustained ventricular tachycardia. A study utilizing programmed stimulation. Am. J. Cardiol. 1075-81, 1982.
18.DiMarco JP, Garan H, Ruskin JN: Mexiletine for refractory ventricular arrhythmias: results using serial electrophsyiologic testing. Am. J. Cardiol. 47: 131-38, 1981.
19.Duff HJ, Rodin D, Primm RK, Oates JA, Woosley RL: Mexiletine in the treatment of resistant ventricular arrhythmias: enhancement of efficacy and reduction of dose related side effects by combination with quinidine. Circ. 67: 1124-28, 1983.
20.Greenspan AM, Spielman SR, Webb CR, Lowenthal DT, Horowitz LN: Efficacy of combination therapy with mexiletine and a type IA agent for sustained ventricular tachyarrhythmias. JACC 3(2): 557, 1984.
21.Zipes DP, Troup,PJ: New antiarrhythmic agents: Amiodarone, Aprinidine, Disopyramide, Ethmozin, Mexiletine, Tocainide, Verapamil. Am. J. Cardiol. 41: 1005-24, 1978.
22.Lalka D, Meyer MB, Duce BR, Elvin AT: Kinetics of the oral antiarrhythmic lignocaine congener, tocainide. Clin. Pharmacol. Ther. 19: 757-66, 1976.
23. Graffner C, Conradson T, Horvendahl S, Ryden L: Tocainide kinetics after intravenous and oral administration in healthy subjects and in patients with acute myocardial infarction. Clin. Pharmacol. Ther. 27: 64-71, 1980.
24. Weigers U: Pharmacokinetics of tocainide in patients with endstage renal failure. In Pottage and Ryden (Eds) Workshop on Tocainide. Proceedings from a conference held in Copenhagen, pp 100-04, 1979. (AB Hassle, Molndal Sweden).

25. Swedberg K, Pehrson J, Ryden I: Electrocardiographic and hemodynamic effects of tocainide (W-36 095) in man. Eur. J. Cardiol. 14: 15-19, 1978.
26. Ikram A: Hemodynamic and electrophysiologic interactions between antiarrhythmic drugs and beta-blockers with special reference to tocainide. Am. Heart J. 100: 1076-80, 1980.
27. Nyquist O, Forssel G, Norlander R, Schenck-Bustafsson K: Hemodynamic and antiarrhythmic effects of tocainide in patients with acute myocardial infarction. Am. Heart J. 100:1000-1005, 1980.
28. Moore EN, Feldman SH, Horowitz LN, Moller RA, Spear JF: Electrophysiologic properties of a new antiarrhythmic drug - tocainide. Am. J. Cardiol. 41: 703-709, 1978.
29. Horowitz LN, Josephson ME, Farshidi A: Human electropharmacology of tocainide, a lignocaine congener. Am. J. Cardiol. 42: 276-80, 1978.
30. Campbell RWF, Bryson LG, Bailey BJ, Muray A, Julian DG: Oral tocainide in suspected acute myocardial infarction. Circ. 60 (Suppl II): 70, 1979.
31. Morganroth J, Panidis IP, Harley S, Johnson J, Smith E, MacVaugh H: Efficacy and safety of intravenous tocainide compared to intravenous lidocaine in patients with acute ventircular arrhythmias immediately after cardiac surgery. Am. J. Cardiol., 1984 in press.
32. Sonhag C: Efficacy and tolerance of tocainide during acute and long-term treatment of chronic ventricular arrhythmias. Eur. J. Cardiol. 18: 301-10, 1980.
33. Winkle RA, Meffin PJ, Fitzgerald JW, Harrison DC. Clinical efficacy and pharmacokinetics of a new orally effective antiarrhythmic, tocainide. Circ. 54: 884-89, 1976.
34. Woolsey RL, McDevitt DG, Nies AS, Smith RF, Wilkinson GR, Oates JA: Suppression of ventricular ectopic depolarizations by tocainide. Circ. 56: 980-85, 1977.
35. Agnew M, Whitlock RML: A double blind crossover study of tocainide, procainamide durules and placebo in patients with chronic ventricular ectopic activity. NZ. Med. J. 91: 363, 1980.
36. Morganroth J, Price B, Dreifus LS: Comparative efficacy and tolerance of quinidine vs. low dose tocainide. Clin. Res. 30: 207A, 1982.
37. Winkle RA, Meffin PJ, Harrison DC: Long-term tocainide therapy for ventricular arrhythmias. Circ. 57: 1008-16, 1978.

38. Ryan W, Engler R, LeWinter M, Karliner JS: Efficacy of a new oral agent (tocainide) in the acute treatment of refractory ventricular arrhythmias. Am. J. Cardiol. 43: 285-91, 1979.
39. Morganroth J: Tocainide: the newest antiarrhythmic agent. In press
40. Kavenina NV, Senova ZP, Mitrofanov VS: Pharmacology of ethomozin: a new antiarrhythmic drug. Farmakol Toksi Kol 35: 182-85, 1972.
41. Ethmozin Investigational Brochure. Endo Laboratories, Inc. pp. 88, December 1981.
42. Danilo P, Langan WB, Rosen MR, Hoffman BF: Effects of the phenothiazine analog EN 313 on ventricular arrhythmias in the dog. Eur. J. Pharmacol. 45: 127-39, 1977.
43. Morganroth J, Michelson EL, Kitchen JG, Dreifus LS: Ethmozin: electrophysiologic effects in man. Am. J. Cardiol. 47(IV): 263, 1981.
44. Mann DE, Luck JC, Pratt CM: Clinical and electrophysiologic effects of ethmozin in man. Clin. Res. 29: 813A, 1981.
45. Morganroth J, Pearlman AS, Dunkman WB. Ethmozin: A new antiarrhythmic agent developed in the USSR. Efficacy and tolerance. Am. Heart J. 98: 621-28, 1979.
46. Kennedy HL, Pescarmona JE, Jayner V, Caralis D, Poblete P. The efficacy and tolerance of ethmozin for chronic ventricular ectopy. Circ. 37(suppl 2): 177, 1978.
47. Podrid PJ, Lyakisheva A, Lown B, Mazur N. Ethmozin, a new antiarrhythmic drug for suppressing ventricular premature complexes. Circ. 61: 450-57, 1980.
48. Pratt CM, Yepsen SC, Addison AT, Mason DT, Miller RR, Quinones MA, Lewis RA. Ethmozin suppression of single and repetitive ventricular premature depolarizations during therpay: documentation of efficacy and long term safety. Am. Heart J. 106: 85-91, 1983.
49. Pratt CM, Young JB, Francis MJ, Taylor AA, Norton HJ, English L, Mann DE, Kopelen H, Quinones MA, Roberts R. Comparative effect of disopyramide and ethmozin in suppressing complex ventricular arrhythmias by use of a double blind, placebo-controlled, longitudinal crossover design. Circ. 69: 288-97, 1984.
50. Podrid PJ, Lown BL: Ethmozin for malignant ventricular arrhythmia. Am. J. Cardiol. 49: 1015, 1982.
51. Morganroth J, Orth DW, Michelson EL, Dunkman WB, Pearlman AS, Horowitz LN, Josephson ME, Kastor JA: Comparative antiarrhythmic efficacy and tolerance of ethmozin and quinidine. Circ. 57 & 58 (II): 103, 1978.

7

CLASS IC ANTIARRHYTHMIC AGENTS: STATUS - 1984

JOEL MORGANROTH, M.D.

With the recognition that patients with chronic non-hemodynamically significant (and potentially lethal) ventricular arrhythmias and left ventricular dysfunction can be identified as being at high risk for sudden cardiac death, protocols have been implemented to determine whether this endpoint can be eliminated. The group of antiarrhythmic agents, which are known as Vaughan Williams Class IC agents have demonstrated particularly unique clinical characteristics that differentiate them from the other Class I drugs agents. Class IC agents have also been shown to be markedly effective in reducing to a very low frequency ventricular arrhythmias in the majority of treated patients. Their electrophysiologic effects predict that they would have very little effect on repolarization and thus the JT interval is only minimally prolonged. They have a marked effect on indicies of depolarization, such that, the QRS interval is often markedly prolonged. Subsequently, prolongation of the QT interval is predominantly due to QRS duration whereas the QT prolongation seen in patients dosed with Class IA drugs (for example, quinidine) is due predominantly to prolongation of the JT interval. Class IC antiarrhythmics have, therefore, been dubbed by some enthusiasts as "PVC killers". Because of their potency, their use in patients with serious underlying left ventricular dysfunction and lethal ventricular arrhythmias can be associated with marked proarrhythmic responses and death if proper care in their use is not exercised. The initiation of dosing with these agents using small dose levels with escalation performed slowly after steady state has been reached has markedly decreased the rate of proarrhythmia and other serious adverse reactions in patients with lethal arrhythmias. Proarrhythmic events using Class IC agents in patients with potentially lethal arrhythmias apears to be low.

The following text will summarize the current status of each of the five Class IC antiarrhythmic agents that are under active investigation in the United States. As of this writing flecainide has been suggested for approval by the Cardio-Renal Advisory Committee and should be available in 1985. New drug applications have either been submitted or will shortly be submitted on encainide, lorcainide and propafenone. Indecainide is undergoing early Class III investigation as of this time.

This potent new class of antiarrhythmic agents which can effectively eliminate ventricular ectopy may be the drug group to use to determine whether sudden cardiac death prevention can be achieved in patients with potentially lethal ventricular arrhythmias using the therapeutic endpoint of arrhythmia control.

FLECAINIDE

Flecainide is a potent IC antiarrhythmic agent synthesized in the United States from a series of a "new breed" of antiarrhythmic drugs of the trifluoroethoxy-benzamide structure. It is a single benzamide with two very distinctive structural features: two trifluoroethoxy groups in the aromatic portion of the molecule and a piperidine ring in the amide side chain. Both of the features are essential to the unique properties of flecainide. Riker Laboratories, Minneapolis, Minn is the sponsor.

MECHANISM OF ACTION

In patients with sinus node dysfunction, Vik-Mo et al[1] found that flecainide effects sinus node function and thus, flecainide must be avoided or sinus node function must be carefully assessed in patients with clinical or laboratory evidence of sinus node dysfunction.

Effects of Conduction

Hellestrand et al[2] assessed the effect of flecainide by electrophysiologic study with plasma levels achieved between 85 and 785 (mean 335) ng/ml. All measured parameters of conduction such as the PA, AH, and HV intervals, QRS duration, QT intervals increased, in contrast, the JT interval signficantly shortened.

Flecainide also produced complete antegrade accessory pathway block in 2 of 8 patients with overt ventricular preexcitation and significantly increased antegrade refractoriness in 6 of 8. Sixteen accessory pathways demonstrated retrograde conduction in the basal state. Seven were concealed, and all 7 of these were completely blocked by flecainide. Right ventricular refractoriness was significantly prolonged from 220 ± 22 to 229 ± 23 msec (mean of 9 msec) and there was no changed in atrial or AV nodal refractoriness.

PHARMACOLOGY

Studies of the hemodynamic effects of acute and chronic oral administration are necessary to ascertain the clinical safety in patients with left ventricular dysfunction since negative inotropy has been detected in some patients especially with pretreated congestive heart failure. Single oral doses ranging from 60 to 240 mg. (0.65-3.53 mg/kg) resulted in peak plasma levels of 69-240 ng/ml proportional to dose, at about 2.5 hours (1.50-3.0 hours, mean 2.6). About 95% of the oral dose of flecainide is absorbed and delivered unchanged to the systemic circulation. The drug is 35 to 42% bound to plasma proteins. Following single doses by either route, the plasma half-life ranges from 7 to 22 hours (mean 14), independent of dose.[3,4] At steady state, the mean elimination half-life is 20 hours, ranging from 12-27 hours.[5,6]

As far as the effects on the ECG are concerned, at doses which demonstrate chronic arrhythmia suppression, flecainide uniformly increases the PR interval (16-37%) and the QRS duration (10-27%)[5,7] One study found a weak correlation between the antiarrhythmic effect and the extent of QRS prolongation (r=0.68), as well as the plasma concentration of flecainide (r=0.70).[7] The lowest plasma level associated with 95% suppression of ventricular extrasystoles on two consecutive days ranged from 217 to 414 ng/ml (mean 317 ng/ml).

EFFICACY

Benign and Potentially Lethal Ventricular Arrhythmias

A multi-institutional, single-blind, placebo control protocol[8] demonstrated 97% VPD suppression in 31 of 33 patients with chronic frequent VPDs by oral flecainide at doses ranging from 100 to 300 mg every 12 hours (mean 206 mg). At least a third of these patients had been resistant or intolerant to multiple antiarrhythmic agents. In all 31 responders during a 2 week follow-up period, there was a sustained 94% suppression rate.

Comparative

A large, double-blind, 16-center trial compared oral flecainide acetate (200-300 mg) twice daily with quinidine sulfate (300-400 q6h) in 280 patients with chronic VPDs.[9] In this parallel study there was an 80% suppression of VPDs in 85% of patients on flecainide, as compared to 57% of patients on quinidine. In addition, flecainide completely suppressed couplets and VT in 68% of patients, compared to 33% of patients completely suppressed in the quinidine group.

Sustained Ventricular Tachycardia

Anderson et al[10] studied 6 patients with inducible ventricular tachycardia lasting at least 30 seconds, or resulting in symptoms or hemodynamic compromise at initial study. Flecainide was administered to each patient at a dose of 150 or 200 mg every 12 hours for 4 to 6 days. A 50% success rate was indicated in this pilot study.

ADVERSE EFFECTS

General

During a multicenter clinical trial[9] of oral administration of flecainide for 2 weeks, 19 of 141 patients (13%) had adverse effects which required cessation of therapy wile, 84 of 141 patients (60%) reported at least one side effect. The most common adverse effects were dizziness (30%) blurred vision (28%), headache (9%) and nausea (9%). In a comparable group of 139 patients receiving quinidine (300 to 400 mg q6h) 15% had side effects requiring discontinuation of therapy while the overall incidence of adverse effects was 65%.

Proarrhythmic Effect

The definition of proarrhythmic effect has been proposed elsewhere.[11] The proarrhythmic effects may be quite different when the patients are separated in to those with lethal ventricular arrhythmias and those with potentially lethal arrhythmias.[11,12]

ENCAINIDE

Encainide is a potent IC antiarrhythmic agent whose chemical synthesis was derived from investigation of a number of open-chain analogs of methysergide in search of potential prophylactic agents for vascular headache. The chemical structure of encainide is 4-methoxy-2'[2-1-methyl-2-piperidyl)ethyl[benzanilide. Bristol-Myers, Evansville, Indiana is the sponsor.

METHODS OF ACTION

Encainide does little to sinus node function in patients without previous evidence of dysfunction.[12,13] However, in patients with prolonged sinus node recovery times at baseline, the drug can result in further dysfunction.

Sami et al[14] found that in 10 patients with coronary artery disease who were given encainide (0.6 to 0.9 mg/kg, by intravenous infusion over a 15 minutes period), the HV and QRS intervals increased by 31 and 18% respectively in proportion to the plasma encainide concentration. There was no significant change in A-H intervals, corrected sinus node recovery time, Wenckebach cycle length, QT interval, or refractory periods of the right atrium, AV node, or right ventricle.

After steady state oral dosing, the electrophysiologic profile of encainide is different than that resulting from acute single-dose intravenous infusion.[13,15] Jackman et al[15] has noted that oral encainide significantly lengthened intra-atrial, A-V nodal, His-Purkinje, and perhaps intraventricular conduction as measured by the PA interval, the AH interval, the shortest atrial pacing cycle length maintaining 1:1 AV nodal conduction, the HV interval and the QRS intervals. In patients with preexcitation syndrome, oral encainide completely blocked antegrade conduction in accessory pathways when the minimum cycle length allowing 1:1 antegrade conduction is greater than 260 msec in the control state.

Thus, the electrophysiologic effects of encainide after steady-state oral dosing: a) are more diverse than predicted by acute intravenous drug dosing, and b) are not correlated ith the serum levels of encainide or its metabolites, or combinations of these levels.

PHARMACOLOGY

The data[16,17] suggest that oral encainide has no clinically significant negative inotropic effect even in patients with or without an impaired left ventricular ejection fraction or a history of dyspnea at rest during short-term follow-up. The mean elimination half-life 3.4 ± 0.3 hours under steady state conditions.[16] Wang et al[18]found bimodal pharmacokinetics among 10 healthy men who received unlabeled encainide, 50 mg every 8h and 54 uCuries of ^{14}C-encainide intravenously concomitantly with the first and eight oral dose. Eight subjects were rapid encainide metabolizers with a mean systemic availability of 26 ± 19%, plasma encainide level of 19 ng/ml and half-life of 2.67 ± 0.93 hr. Two subjects were slow metabolizers with a systemic availability of 89 ± 13%, serum encainide levels which are 20-fold higher (mean 377 ng/ml), and a mean half-life of 8.7 ± 3.4 hr. Rapid metabolizers produced 3-methoxy-O-demthylencainide (MODE). In contrast, the two slow metabolizers produced no detectable MODE but made larger amounts of O-demethylencainide (ODE). In neither group did unmetabolized encainide accumulate significantly. Both major metabolities, (ODE), and (MODE), are pharmacologically active and have longer plasma half-lives than the parent compound.[19-24] The elimination half-life of ODE was 11 ± 10 hours. The plasma half-life of MODE could not be calculated but is probably about 1 day.[16] The duration of action, as measured by 90% ventricular premature depolarization suppression after a single intravenous dose, ranged from 1 to 36 hours with a mean and median of 8 hours, and ranged from 3 to 25 hours after a single oral dose.[25]

EFFICACY

Benign and In Potentially Lethal Ventricular Complexes

Roden and coworkers[16] in a dose ranging, double-blind, oral protocol which included both placebo control and washout periods found almost complete abolition of ventricular ectopy (>99% reduction) in 10 of 11 patients, 7 of whom had been resistant or intolerant to 4 or more previous drug regimens. The doses ranged from 25 to 100 mg every eight hours, and efficacy persisted for 6 to 12 months of follow-up.

DiBianco et al[13] found that oral encainide (mean dose, 75 mg, every 6 hours), titrated to Holter-monitor response resulted in an 80% reduction in ventricular premature depolarization frequency in 20 of 21 patients with ventricular premature complexes (VPC) frequency of $\geq$ 60 beats/hour. There was a reduction in couplets by 99.8% and in salvos of nonsustained ventricular tachycardia by 99%.

Morganroth and coworkers[26] found that encainide, (25-75 mg every 6 hours), significantly reduced the VPC frequency in 27 of 30 patients, with complete suppression (<1 VPC/hr/24hrs) in 13 (48%) of these patients. Complex VPCs were abolished in 16 (84%) of 19 patients. At a dosage $\leq$ 50 mg, every 6 hours, which was devoid of any side effects, VPCs were abolished in 10 (77%) of 13 patients, and complex VPCs were abolished in 13 (81%) of 16 patients.

Sami et al[27] conducted fixed oral quinidine vs encainide study in 20 men with VPCs. Encainide (50 mg, every six hours) suppressed all VPCs recorded on ambulatory electrocardiogram in 44% of patients, reduced the frequency by 80% or more in 88% of patients, and suppressed complex forms in all patients. On the other hand, quinidine (300 mg, every six hours) failed to totally suppress VPCs in any patient, resulted in 80% reduction in VPC frequency in less than half of the patients, and suppressed complex forms in half of the patients. Encainide was also better tolerated than quinidine.

In Refractory Recurrent Sustained Ventricular Tachycardia

The use of encainide in lethal ventricular arrhythmias has not been well reported in any study using current techniques. An evaluation of the data by Dr. Morrison Hodges for Bristol-Myers Company has revealed the encainide prevented re-induction of sustained ventricular tachycardia in 6/15 (48%) of patients tested.

SIDE EFFECTS

Encainide, administered chronically to 119 patients demonstrated the most commonly produced side effect to be blurring of vision (23 patients - 19%). Dizziness or ataxia occurred in 14 patients (12%). Headache, tremor, nausea, and constipation were seen in <3%. In only one patient with preexisting visual impairment, was encainide discontinued because of side effects. Occurrence of adverse effects appears to be determined by the total daily dose and the peak plasma level attained. The incidence of side effects in 67% at doses greater than 50 mg every six hours, with only a 14% incidence of side effects at lower doses.[26] Asymptomatic sinus pauses (less than 2 sec. duration) have also been reported to occur in 3% of cases.[13] There has been one reported cases of development of third-degree atrioventricular block in a patient on encainide therapy. However, this patient presented with pronounced QRS widening before receiving the drug.[28]

PROARRHYTHMIC EFFECTS

Proarrhythmic effects were reported in 10 (8%) of 119 patients on chronic oral encainide.[12,13,16,26,27] Half of these had a previous history of sustained ventricular tachycardia requiring cardioversion.

Like the neurologic and gastrointestinal side effects, there is evidence that the proarrhythmic effects of encainide are dose related.[28]

Although proarrhythmic effects appear to be dose-related, they cannot be predicted by QRS interval prolongation which is usually moderate, or QTc interval prolongation, which is usually minimal. Other than a history of previous malignant arrhythmias, the only other factor which seemed preditive of proarrhythmic effect was the temporal relationship of less than 48 hours after initiation or oral encainide therapy.

Although proarrhythmic effects appear to be dose-related, they cannot be predicted by QRS interval prolongation which is moderate, or QTc interval prolongation, which is usually minimal. Other than a history of previous malignant arrhythmias, the only other factor which seemed preditive of proarrhythmic effect was the temporal relationship of less than 48 hours after initiation or oral encainide therapy.

INTERACTIONS WITH OTHER CARDIOVASCULAR DRUGS

Among 10 patients who were on chronic maintenance doses of digoxin, 1 to 2 weeks of therapy with encainide resulted in no significant change in the mean serum digoxin level (1.04 $\pm$ 0.43 before, 1.22 $\pm$ 0.47 after).[17]

LORCAINIDE

Lorcainide is a potent type IC antiarrhythmic drug with local anesthetic activity. Its chemical names is N-(4-chlorophenyl)-N-[1-(1-methylethyl)-4-piperldinyl]benzene acetamide monohydrochloride. The sponsor is Janssen Pharmaceutica, New Brunswick, N.J.

MECHANISM OF ACTION

In patients with pre-existing sinus node dysfunction, lorcainide must be used with great caution unless pacing is available. Manz et al[29] noted that after lorcainide administration (2 mg/kg intravenously) 3 of 7 patients developed heart rates of less than 50 beats/min due to sinus node exit block. There was marked prolongation of the corrected sinus node recovery time.

Lorcainide tends to prolong the atrial effective refractory period while most studies have shown no changes in the right ventricular refractory period as well as the effective refractory period of the AV node. In patients with Wolff-Parkinson-White syndrome both retrograde and anterograde refractory periods of accessory pathways were prolonged by lorcainide.[29,30,31,33,34]

The effects of lorcainide on the surface electrocardiogram consists of prolongation of QRS duration[32,35,36] as well as of P-wave duration, PR and QTc interval[35,37] although other investigators have not confirmed these results.[38]

Oral treatment with lorcainide was found to increase A-H interval and ventricular effective refractory period[39,40] while it did not alter sinus cycle length.[40]

PHARMACOLOGY

The effects of lorcainide on myocardial function has been studied by several investigators.[41,42] It appears then that lorcainide exerts a moderate, although, probably transient, myocardial depressant effect and should be administered with caution in patients with underlying left ventricular dysfunction. A steady-state, bioavailibility is approximately 100%. Orally administered lorcainide is metabolized to a pharmacologically active compound, norlorcainide. Metabolities may play an additional role in arrhythmia suppression.[35,39] The elimination half-life of lorcainide was found to be close to 7.7 hours by some investigators[43-45] while Somani[46] found a longer elimination half-life of 13.1 hours. These different findings may be explained by interindivididual variability in lorcainide pharmacokinetic variables. Because lorcainide is virtually completely eliminated by hepatic biotransformation with a high hepatic extraction ratio, liver diseases, such as cirrhosis[47] and congestive heart failure[48], significantly prolong the mean elimination half-life and produce a reduction in clearance. Klotz et al[48] found the plasma concentration of lorcainide, associated with effective suppression of ventricular ectopic beats, to be 150 to 400 ug/L. Since lorcainide has a wide therapeutic range, individualization of therapy is necessary. However, total daily doses >600 mg have occasionally been used.[50]

EFFICACY

Benign and Potentially Lethal Ventricular Ectopic Beats

Several open studies have shown a favorable effect of lorcainide on VPCs. A >80% suppression of VPCs was found in 23 of 27 patients treated with an average dose of lorcainide of 139 mg intravenously.[52] Furthermore, Meinertz et al[50] noted suppression of these arrhythmias in 11 of 12 patients treated with oral therapy. Included were 7 of 8 patients who failed to respond to at least one antiarrhythmic agent. Others[35,37,46,49,51,53,54] have demonstrated similar long-term efficacy. Amery et al[55] in a review of lorcainide of numerous studies noted a positive response (defined as >50% reduction of VPCs) in 84% of patients treated orally with the drug and in 82% treated intravenously.

Morganroth et al[56] in a single-blind study, treated 12 patients with lorcainide 100 mg twice daily and reported that 8 of them had >70% suppression of VPCs.

Vlay et al[57] have studied 13 patients using a crossover study or lorcainide (100 bid to tid) and quinidine gluconate 324 mg tid. Eight of 13 patients on lorcainide compared to 3/13 on quinaglute had a >80% reduction of potentially lethal or benign VPCs.

In Sustained Ventricular Tachycardia

Lorcainide, intravenously or orally, has eliminated inducible ventricular tachycardia in a variable proportion of patients ranging from 0 to 100%. Sung and Somani[58] studied 5 patients treated with 2 mg/kg of intravenous lorcainide and all of them had induction of ventricular tachycardia despite treatment. Somberg et al[59] reported suppression of ventricular tachycardia in all 7 patients treated with 100 mg of intravenous lorcainide. The drug was, however, administered only 9 hours after intravenous procainamide and plasma concentration of either drug were not reported. Several studies reported a 36 to 45% of patients in whom oral lorcainide produced suppression of ventricular tachycardia induced by programmed electrical stimulation or exercise testing.[40,60]

SIDE EFFECTS

Although side effects are common during oral lorcainide therapy, they are rarely serious in nature and tend to resolve or diminish with continuation of treatment. The most common side effects (nearly 100% of patients) are sleep disturbances, which usually resolve with time if lorcainide is continued, and usually responds well to a benzodiazepine.[56] Other common side effects include headache, nausea, excessive perspiration and hot sensation and, during intravenous administration, a tingling sensation. Proarrhythmia has been noted.[60]

PROPAFENONE

Propafenone is a potent Class IC antiarrhythmic class and was synthesized in 1970 and commercially available in Germany since 1977. The chemical name for propafenone is 2'_[3-(prophylamino)-2-(hydroxy)-(propoxy)-3-phenylpropiophenone hydrochloride). The sponsor is Knoll Pharmaceutical, Whippany, N.J.

MECHANISMS OF ACTION

Propafenone[61] can reduce sinus node function significantly. Significant bradycardia was seen in patients with sinus node dysfunction. It significantly increases the PQ time, the AH interval, HV interval and the width of the QRS complex. Propafenone can cause a moderate decrease in contractility without altering either the heart rate or the aortic pressures.

PHARMACOLOGY

Mild-to-moderate depression in ventricular function has been observed in clinical studies.[62,63]

Propafenone is probably extensively metabolized most likely by first-pass hepatic metabolism. The metabolic pathway has not been completely elucidated. The comparison of the cumulative urinary excretion over 24 hours showed at 1.3% of the intravenous (70 mg) and 0.65% of the oral substance (600 mg) were excreted in the urine in unchanged form. Connolly and coworkers[64] reported a great intersubject variability in elimination of half-life from 2.4 to 11.8 hours.

Pronounced interindividual variations were found for frequency of VPCs correlated to plasma concentration. These variations were even more pronounced during intravenous therapy than during oral medication. Wiebringhaus et al[65] studied 43 patients and found the daily doses between 450 and 1200 mg and the plasma concentrations between 0.5 and 3.0 ug/ml. There was no strict correlation between plasma levels and dose ingested, therapeutic success and side effects. However, there was some correlation between plasma levels and widening of the QRS complex. Connolly et al[64] reported that the plasma propafenone concentration that caused over 90% suppression of VPCs ranged from 64 to 1,044 ng/ml (mean 588 ng/ml).

EFFICACY

Benign and Potentially Lethal Ventricular Arrhythmias

In a double-blind, crossover trial, in which propafenone was compared against placebo, DeSoyza et al[66] reported propafenone to be effective in 25 out of 30 patients. Ventricular tachycardia, which was present in 29 patients at baseline, was abolished in all patients at optimum propafenone dosages (about 900 mg/day).

Salerno et al[67] using Holter recordings to monitor efficacy in a double-blind, crossover, placebo controlled study, evaluated the effectiveness of propafenone for the treatment of VPCs in 12 patients. The effectiveness of propafenone was confirmed: 80% suppression of total VPCs in 7 of 12 patients and almost complete suppression of ventricular tachycardia. In another study 10 patients with VPCs, 24-hour Holter monitoring was performed before and after administration of propafenone (900 mg/day) and quinidine (800 mg/day). Propafenone completely controlled the arrhythmia in 8 patients, quinidine in only 3.[68]

Naccarella and coworkers[69] compared the efficacy of propafenone (300 mg tid for 5 days) with disopyramide (200 mg tid for 5 days) using 24 hour Holter monitoring in 12 patients with symptomatic VPCs. The results indicated that both drugs significantly reduced extrasystole's rates and clinical effectiveness (70-80% reduction in total ventricular extrasystole) was obtained in 8 of 12 patients receiving propafenone and 6 of 12 receiving disopyramide. Klempt et al[70] evaluated the efficacy of mexiletine (600 mg/day), propafenone (900 mg/day) and flecainide (400 mg/day) in 12 patients following a myocardial infarction. The reduction in single VPCs >90% were as follow: flecainide (10/12 patients), propafenone (8/12 patients) and mexiletine (4/12 patients).

In Lethal Ventricular Arrhythmias

Programmed electrical stimulation studies were performed both before and after oral (900 mg/day) propafenone.[71] Ventricular tachycardia cycle length increased significantly. Following oral evaluation, 4 of 8 patients were discharged on propafenone and had no symptomatic tachycardia or side effects with a mean follow-up period of 4 months. Connolly et al[77] performed electrophysiologic studies in 16 patients with ventricular tachycardia or non-fatal cardiac arrest. All patients had baseline inducible sustained ventricular tachycardia. During oral propafenone therapy (150 mg to 300 mg, three times daily) ventricular tachycardia was not inducible in 1 patient, was unsustained in 1 and was harder to induce in 2 patients. Two of 5 patients, treated with oral propafenone, were arrhythmia-free during a follow-up period of 1 to 8 months.

SIDE EFFECTS

Side effects include restlessness, disturbed sleep, lack of concentration, syncope, myoclonus, psychosis, convulsion, diarrhea, epigastric pain, urticaria and alopecia. Rare cases of cholestatic hepatitis and reversible impotence have been reported. In Europe, two cases of cholestatic hepatitis have been related to propafenone treatment.[73] Experience with long-term treatment has shown that extra cardiac side effects may still occur after considerable time.[74] Proarrhythmia can occur.

DRUG INTERACTION

In a study Belz and co-workers[75] it was indicated that propafenone (150 mg q8h) increased plasma digoxin concentrations 37% during co-administration for 2 weeks. In another study, propafenone increased serum digoxin levels in 5 of 5 patients (mean increase of 83%).[67] However, no patients had evidence of digitalis toxicity during this study.

INDECAINIDE

HISTORY

Indecainide (LY135837) is a new investigational antiarrhythmic agent, developed through a systemic evaluation of antiarrhythmic agents by Eli Lilly and Company, Indianapolis, Indiana. The drug is a type IC antiarrhythmic agent. Its chemical name is 9-[3-[(Methyl ethly) amino]propyl]-9H-Fluorene-9-carboxamide, hydrochloride.

MECHANISM OF ACTION

Indecainide, in humans, prolongs cardiac impulse conduction expressed in the surface electrocardigoram as lengthening of the PR and QRS intervals. Indecainide was given IV to 10 patients with coronary artery disease and sustained ventricular tachycardia and was found to elevate the AH, HV, and QRS with minimal prolongation of the JT interval.[76] Siddoway et al,[77] in a study of 10 patients treated with oral indecainide (50-100 mg, every 8 hours), noted that the PR interval increased 0.4 ± 0.02 sec. in responders ($p<0.5$) and only 0.1 sec. in non-responders. Krejci and coworkers[78] reported a 25% increase in the PR interval and an 18% increase in the QRS interval while the QT interval nonsignificantly decreased by 1%.

PHARMACOLOGY

In one study, 5 subjects were treated with indecainide (doses varied from 1.5 yg/kg/min for 25 min to a maximum of 25 ug/kg/min for 17 min). Blood pressures, pulse rate, systolic time intervals, stroke volumes, cardiac output and cardiac index were measured by noninvasive techniques every 5 minutes before, during, and after the infusion of the drug or placebo. Increasing doses of drug were associated with increase in diastolic and mean pulse pressure, pre-ejection period, and heart rate and with decrese in stroke volume, cardiac output and cardiac index.

The negative inotropic effects of the drug are not yet fully studied. However, caution should be exercised when administering indecainide to patients with compromise cardiac function, and when administered concomitantly with other cardiac depressants.

In one study, intravenous administration of ^{14}C radiolabeled indecainde in 4 normal volunteers resulted in the plasma half-life of 5.63 ± 1.19 hours (range 4.8 to 7.9 hours).. Eight-four percent of the drug was excreted in the urine while 11% in the feces. Oral radiolabeled indecainide given to 5 subjects (doses of 0.56 to 0.73 mg/kg) resulted in the plasma half-life of 11.59 ± 5.78 hours (range 6 to 12 hours). Eight-one percent of the radioactivity was excreted in the urine and 17% in the feces.[79] Thus, it is possible that 17% of the oral administered drug undergoes first-pass metabolism through liver. Approximately 87% of the oral dose is absorbed through the intestinal tract. Morganroth et al[79] demonstrated that the metabolic desisopropyl indecainide represented 10% of the active compound.

In one study of oral indecainide in 5 subjects, the maximum plasma concentrations ranged from 152 to 315 ng/ml (mean 217.4 ± 72.9 ng/ml), and was reached in 1.9 ± 0.9 hours (range 0.5 to 25 hours). Krejci et al[78] reported that indecainide plasma levels, on optimal dose, were 233-658 ng/ml (mean 413 ng/ml).

EFFICACY

In Benign Potentially Lethal Ventricular Arrhythmias

In a dose ranging study, intravenous indecainide (17.5 ug/kg/min for 25 min) produced a 100% suppression of the ventricular ectopic activity for a short period of time. When 60 mg were infused at 12.6 mg/kg/min, ventricular ectopic beats were completely suppressed for 10 to 11 hours. Oral indecainide was administered to patients with ventricular arrhythmias. Initially, the drug was given in single doses of 25 to 90 mg in 3 patients and the agent produced a decresed in ventricular ectopic activity during the first control period. A second control period followed with indecainide given at 50 mg three times a day initially and then four times daily. Considerable deduction (85-90%) of ventricular ectopic beats occurred during drug therapy. In the final phase of the study indecainide at doses of 180-200 mg/day suppressed the ventricular ectopic beats for a period of 20 days while paired beats were reduced only 50-60%.

Morganroth et al[79] reported a >70% reduction in VPCs in 93% of patients and >90% reduction in 85% of patients, ventricular tachycardia was abolished in 85%. In another study, during dose ranging, all patients achieved at least 80% reduction of total ventricular ectopic depolarization. Eight of 10 patients, studied by Siddoway et al[77] had satisfactory efficacy at a mean dosage of 66 ± 19 mg every 8 hours.

In Ventricular Tachycardia

In 10 patients with coronary artery disease and sustained ventricular tachycardia evaluted with electrophysiology testing, indecainide was given at 60-90 mg intravenously at 12.5-15 ng/kg/ in. Countershock was required before indecainide in 7/10 patients and afterwards in only 3/10 patients. Only 1 patient was not inducible. The mean plasma level was 553 ng/ml (193-876).[80]

Side Effects

Siddoway and coworkers[77] reported side effects in 3 of 10 patients and consisted of liver transaminase elevated (1 patient), dizziness (2 patients) and decrease in white blood count from 4880 ± 283 to 3250 ± 212 (2 patients). Changes in white blood count was not observed in later studies. Krejci et al[78] noted side effects in 4 of 7 patients but none of them required cessation of the drug. Three patients experienced dizziness, one patient developed asymptomatic postural hypotension, one patient developed right bundle branch block and another one complained of blurred vision. In the study by Morganroth et al,[80] 2 of 14 patients reported side effects (blurred vision, and parasthesia in 1 each). Other adverse effects reported include headache, nausea, vomiting, a "funny taste" and drowsiness.

REFERENCES

1. Vik-Mo H, Ohm O, Lund-Johansen P: Electrophysiologic effects of flecainide acetate in patients with sinus node dysfunction. Am J Cardiol 50:1090-1094, 1982.

2. Hellestrand KJ, Bexton RS, Nathan AW, Spurrel RAJ, Acamm AJ: Acute electrophysiologic effects of flecainide acetate on cardiac conduction and refractoriness in man. Br Heart J 48:140-148, 1982.

3. Conard CJ, Carlson GL, Frost JW, Ober RE: Human plasma pharmacokinetics of flecainide acetate (R-818), a new antiarrhythmic, following single oral and intravenous doses. Clin Pharmacol Ther 25:218, 1979. (Abstract)

4. Conard CJ, Carlson GL, Frost JW, Ober RE: Human plasma elimination kinetics of R-818, a new antiarrhythmic. In Abstracts of Papers Presented before the APHA Academy of Pharmaceuticals Sciences, 21st National Meeting. Orlando, p. 106, 1976. (Abstract)

5. Hodges M, Haughland M, Granrud G, Conrad J, Asinger RW, Mikell L, Krejci J: Suppression of ventricular ectopic depolarization by flecainide acetate, a new antiarrhythmic agent. Circulation 65:879-885, 1982.

6. Anderson JL, Stuart JR, Perry B, Van Lammersfeld DD, Johnson RA, Pritt B: Oral flecainide acetate for elimination of ventricular arrhythmias. N Engl J Med 305:473-477, 1981.

7. Duff HJ, Roden DM, Maffucci RJ, Vesper BS, Conard GJ, Higgins SB, Oates JA, Smith RF, Woosely RL: Suppression of resistant ventricular arrhythmias by twice daily dosing with flecainide. Am J Cardiol 48:1133-1140, 1981.

8. Conrad GJ, Cronheim GE, Klempt HW: Relationship between plasma concentrations and suppression of ventricular extrasystoles by flecainide acetate (R-818), a new antiarrhythmic, in patients. Arzeimittel-Forschung 32:155-159, 1982.

9. The Flecainide-Quinidine Research Group. Flecainide versus Quinidine for Treatment of Chronic Ventricular Arrhythmias. A multicenter clinical trial. Circulation 67:1117-1123, 1983.

10. Anderson JL, Lutz JR, Allison SB: Electrophysiologic and antiarrhythmic effects of oral flecainide in patients with inducible ventricular tachycardia. JACC 2:105-114, 1983.

11. Morganroth J, Horowitz LN: Flecainide: Its proarrhythmic effect and expected changes on the surface electrocardiogram. Am J Cardiol 53:989-948, 1984.

12. Morganroth J: Premature ventricular complexes: Diagnosis and Indications for Therapy JAMA 252:673-676, 1984.

13. DiBianco R, Flectcher RD, Cohen AI, Gottdiener JS, Singh SN, Katz RJ, Baes HR, Sauerbrunn B: Treatment of frequent ventricular arrhythmia with Encainide: Assessment using serial ambulaory electrocardiograms, intracardiac electrophysiology studies, treadmill exercise tests, and radionuclide cineangiographic studies. Circulation 65:1134-1147, 1982.

14. Sami M, Mason JW, Peters F, Harrison DC: Clinical electrophysiologic effects on Encainide, a newly developed antiarrhythmic agent. Am J Cardion 44:526-532, 1979.

15. Jackman WM, Zipes DP, Naccarelli GV, Rinkenberger RL, Heger JJ, Prystowsky EN: Electrophysiology of oral encainide. Am J Cardiol 49:1270-1278, 1982.

16. Roden DM, Reele SB, Higgins SB, Mayol RF, Gammans RE, Oates JA, Woosley RL: Total suppression of ventricular arrhythmias by Encainide. Pharmacokinetic and electrocardiographic characteristics. N Engl J Med 302:877-882, 1980.

17. Sami MH, Derbekyan VA, Lisbona R: Hemodynamic effects of Encainide patients with ventricular arrhythmia and poor ventricular function. Am J cardiol 52:507-512, 1983.

18. Wang T, Roden DM, Wolfenden HT, Woosley RL, Wilkinson GR, Wood AJJ: Pharmacokinetics of Encainide and its metabolites in man. (Abstract) Clin Pharmacol Ther 31:278, 1982.

19. Roden DM, Duff HJ, Altenberg D, Woosley RL: Antiarrhythmic activity of the O-Demethyl metabolite of Encainide. J Pharmacol Exp Ther 221:552-557, 1982.

20. Elharrar V, Zipes DP: Effects of Encainide and metabolites (MJ 14030 and MJ9444) on canine cardiac purkinje and ventricular fibers. J Pharmacol Exp Ther 220:440-447, 1982.

21. Kates RE, Harison DC, Winkle RA: Metabolie cumulation during long-term oral Encainide administration. Clin Pharmacol Ther 31:427-432, 1982.

22. Winkle RA, Peters F, Kates RE, Harrison DC: Possible contribution of Encainide metabolites to the long-term antiarrhythmic efficacy of Encainide. Am J Cardiol 51:1182-1188, 1983.

23. Duff HJ, Dawson AK, Roden DM, Oaes JA, Smith RF, Woosley RL: Electrophysiologic actions of O-Demethyl Encainide: An active metabolite. Circulation 68:2, 385-391, 1983.

24. Kate RE, Woosley RL, Harrison DC: Clinical importance of metabolites of antiarhythmic drugs. Am J Cardiol 53:248-251, 1984.

25. Winkle RA, Peters F, Kates RE, Tucker C, Harrison DC: Clinical pharmacology and antiarrhythmic efficacy of Encainide in patients with chronic ventricular arrhythmias. Circulation 64:290-296, 1981.

26. Morganroth J, Hodges M, Ruffy R, Granrud G, Wouthworth W: The Cooperative Encainide Trial: Efficacy, tolerance and dose findings in chronic severe ventricular ectopy. (Abstract) Am J Cardiol 49:1003, 1982.

27. Sami M, Harrison DC, Kreamemer H, Houston N, Shimasaki CN, DeBusk RF: Antiarrhythmic efficacy of Encainide and Quinidine: Validation of a model for drug assessment. Am J Cardiol 48:147-156, 1981.

28. Chesnie B, Podrid P, Lown B, Raeder E: Encainide for refractory ventricular tachyarrhythmia. Am J Cardiol 495-500, 1983.

29. Manz M, Steinbeck G, Luderitz B: Electrophysiological effects of lorcainide in sinoatrial disease and in Wolff-Parkinson-White Syndrome. Eur Heart J 3:56-66, 1982.

30. Bar FW, Farre J, Ross D, Vanagt EJ, Gorgels AP, and Wellens HJJ: Electrophysiological effects of lorcainide, a new antiarrhythmic drug. Observations in patients with and without pre-excitation. Br Heart J 45:292-298, 1981.

31. Kasper W, Meinertz T, Kersting F, Lollgen H, Lang K and Just H: Electrophysiological actional of lorcainide in patients with cardiac disease. J Cariovasc Pharmacol 1:343-352, 1979.

32. Stroobandt R, and Kesteloot H: Lorcainide: A comparison with aprindine; in Jahnchen et al. (Eds) Prognosis and Pharmaco-therapy of life-threathening arrhythmias. pp.129-133, Royal Society of Medicine, London, 1981.

33. Brugada P, Bar FWHM, Friedman PL and Wellens HJJ: Electrophysiological effects of lorcainide: Observations in patients with an accessory pathway: in Jahnchen et al (Eds) Prognosis and Pharmacotherapy of life-threathening arrhythmias. pp. 115-119, Royal Society of Medicine, London, 1981.

34. Kasper W, Treese N, Meinertz T, Jahnchen E, and Pop T: Electrophysiologic effects of lorcainide on the accessory pathway in the Wolff-Parkinson-White Syndrome. Am J Cardiol 51:1618-1622, 1983.

35. Meinertz T, Kasper W, Kersting F, et al: Lorcainide II: Plasma concentration-effect relationship. Clin Pharmaco and Ther 26:196-204, 1979.

36. Ng CK, Gstottner M, and Gneiner R: Intracardiac electrophysiologic effects of lorcainide in man. Eur J Clin Pharmaco 15:241-247, 1979.

37. Myburgh DP, Goldman AP and Schamroth JM: Lorcainide - An antiarrhythmic agent for ventricular arrhythmias. SA Med J 57:236-239, 1980.

38. Keefe DL, Peters F, and Winkle RA: Randomized double-blind placebo controlled crossover trial documenting oral lorcainide efficacy in suppression of symptomatic ventricular tachyarrhythmias. Am Heart J 103:511-518, 1982.

39. Echt DS, Mitchell LD, Kates RE and Winkle RA: Comparison of the electrophysiologic effects of intravenous and oral lorcainide in patients with recurrent ventricular tachycardia. Circulation 68:392-399, 1983.

40. Saksena S, Rothbart ST, Cappello G, Bernstein A, and Somanig P: Clinical and electrophysiologic effects of chronic lorcainide therapy in refractory ventricular tachycardia. J Am Coll Cardiol 2:538-544, 1983.

41. Shita A, Bernard R, Mostinck R, and Debacker M: Haemodynamic reactions after intravenous injection of lorcainide hydrochloride in acute myocardial infarction. Eur J Cardiol 12:237-242, 1981.

42. Meinertz T, Kersting F, Kasper W, Just H, Bechtold H and Jahnchen E: Haemodynamic effect of a single intravenous dose or lorcainide in patients with heart disease. Eur J Clin Pharmacol 18:461-465, 1980.

43. Jahnchan E, Bechtold H, Kasper W, Kersting F, Just H, Heykants J and Meinertz T: Lorcainide I: Saturable presystemic elimination. Clin Pharmaco and Ther 26:187-195, 1979.

44. Klotz U, Muller-Seydiltz P, and Heimburg P: Pharmacokinetics of lorcainide in man: A new antiarrhythmic agent. Clin Pharmaco 3:407-418, 1978.

45. Kates RE, Keefe DL, and Winkle RA: Lorcainide disposition kinetics in arrhythmia ptients. Clin Pharmacol and Ther. 33:28-34, 1983.

46. Somani P: Pharmacokinetics of lorcainide, a new antiarrhythmic drug in ptients with cardiac rhythm disorders. Am J Cardiol 48:157-163, 1981.

47. Koltz U, Fischer C, Muller-Seydlitz P, Schulz J and Muller WA: Alterations in the disposition of differently cleared drugs in patients with cirrhosis. Clin Pharmacol and Ther 26:221-227, 1979.

48. Bodigheimer K: Lorcainide: A comparison with lidocaine in the treatment of ventrticular ectopic beats of various genesis in Jahnchen et al (Eds) Prognosis and Pharmacoltherapy of Life-threatening Arrhythmias, pp.135-140, Royal Society of Medicine, London, 1981.

49. Klotz U, Muller-Seydlitz P, and Heimburg P: Lorcainide infusion in the treatment of ventricular premature beats (VBP). Eur J Clin Pharmacol 16:1-6, 1979.

50. Meinertz T, Kasper W, Kersting F, Bechtold J, Just H, and Jahnchen E: Antiarrhythmic effect of lorcainide during chronic treatment. Arzneimittel-Forschung 30;1593-1595, 1980.

51. Klotz U: Disposition and antiarrhythmic effect of lorcainide. Inter J Clin Pharmaco Bio 17:152-158, 1979c.

52. Kesteloot H, and Stroobandt R: Clinical experience with lorcainide (R15889) a new antiarrhythmic drug. Arch Int Pharmaco Ther 230:225-234, 1977.

53. Terryn R, Keulen UWBM and Amery W: Control of ventricular ectopic beats with intravenous lorcainide maintained by oral doses. Int J Cardiol 3:383-384, 1983.

54. Cocco G, Strozzi C: Initial clinical experience of lorcainide (Ro13-1042) a new antiarrhythmic agent. Eur J Clin Pharmacco 14:105-109, 1978.

55. Amery WK, Heykants JJP, Xhonneux R, Towse G, Oettel P, Gough DA, and Janssen PAJ: Lorcainide (R15889), a first review Acta Cardiol 36:207-234, 1981.

56. Morganroth J: The efficacy and safety of lorcainide in the treatment of ventricular arrhythmias as assessed by frequent Holter monitoring in Jahnchen et al (Eds) Prognosis and Pharmacotherapy of Life-threatening Arrhythmias, pp.149-151, Royal Society of Medicine, London, 1981.

57. Vlay SC, Mallis GI, Singh S and Cohn PF: Comparison of lorcainide and quinidine in the treatment of ventricular ectopy. Chest 86:80-83, 1984.

58. Sung RJ and Somani P: Discordant antiarrhythmic effects of intravenous lorcainide on ventricular premature beats and ventricular tachycardia. Circulation 64:IV-37, 1981.

59. Somberg JC, Willens H, Camilleri W, Maquire W, and Miura DS: Effect of lorcainide on suppressing ventricular tachycardia induced by programmed stimulation 54th Scientific Session of AHA, Dallas, November 16-19, 1981. Circulation 64:IV-37, 1981.

60. Chesnie B, Lampert S, Podrid P, and Lown B: Lorcainide for refractory ventricular arrhythmia. Abstract of the 32nd Annual Scientific Session. American College of Cardiology, New Orleans, Louisiana, March 20-24, 1983. J Am Coll Cardiol 1:708, 1983.

61. Seipel L, Breithardt G, Both A, Wiebringhaus E: Effects of propafenone on the sinus node and intracardiac conduction in man (Ger) Drug Dev Eval 1:45-55, 1977.

62. Beck OA, Vogt J, Hochrein H: The comparative hemodynamic effects of antiarrhythmic drugs in acute myocardial infarction: Comparison of propafenone and lidocaine. Z Kardiol 67:41-45, 1978.

63. Baker BJ, DeSoyza N, Boyd CM, Murphy ML: Effects of propafenone on left ventricular function. Circulation 67:II267, 1982.

64. Connolly SJ, Kates RE, Lebsack CS, Harrison DC, Winkle RA: Clinical pharmacology of proprafenone. Circulation 68:589-596, 1983.

65. Blanke H, Aschbrenner B, Karsch KR, Dreuzer H: Plasma level, effectiveness, and organ distribution of propafenone. Dtsch Med Wochenschr 104:587-591, 1979.

66. DeSoyza N, Murphy M, Sakhaii M, Treat L: The safety and efficacy of propafenone in suppressing ventricular ectopy. Circulation II67:266, 1982.

67. Salerno DM, Granrud G, Sharkey P, Asinger R, Hodges M: A controlled trial of propafenone for treatment of frequent and repetitive ventricular premature complexes. Am J Cardiol 53:77-83, 1984.

68. Barros RB, et al: Propafenone: A new chance for the treatment of ventricular arrhythmias (Ger) 33-Congr Brasil Cardiol, Porto Alegre, July 10-16, 1977.

69. Naccarella F, Palmieri M, Del Corso G, Lombardi G, Bracchetti D: Comparison of the efficacy of propafenone versus disopyramide for complex premature ventricular contractions. Drug Res 32:547-550, 1982.

70. Klempt HW, Nayebagha A, Fabry E: Antiarrhythmic effect of mexiletine, propafenone, and flecainide with ventricular extrasystoles. A comparison study in patients following myocardial infarction (Ger). Ziet fur Kardiol 71:340-349, 1982.

71. Chilson DA, Zipes DP, Heger JJ, Brown KF, Lloyd EA, Prystowsky EN: Propafenone: Clinical and electrophysiologic effects in patients with ventricular tachycardia. Clin Res 30:706A, (Abstract) 1982.

72. Connolly SJ, Kates RE, Lebsack CS, Echt DS, Mason JW, Winkle RA: Clinical efficacy and electrophysiology of oral propafenone for ventricular tachycardia. Am J Cardiol 52:1208-1213, 1983.

73. Schuff-Werner P, Kaiser D: Choolestatic hepatitis following antiarrhythmic therapy with propafenone (Ger) Dtsch Med Wschr 105:137-138, 1980.

74. Fischer G, Seipel L: Clinical experience with oral propafenone (Ger). Drug Dev Eval 1:109-113, 1977.

75. Belz GG, Matthews J, Sel M, Doering MD, Belz G: Digoxin-antiarrhythmias, pharmacodynamics and pharmacokinetic studies with quinidine, propafenone and verapamil. Clin Phar Ther 31:202-203, 1982 (abs).

76. Horowitz LN, Morganroth J, Webb CR, Kay H: Clinical electrophysiology of Indecainide. Circulation in press, 1984. (Abstract)

77. Siddoway LA, Bergstrand RH, Primm KR, Roden DM, Woosley RL: Antiarrhythmic efficacy, safety and pharmacokinetics of indecainide. Circulation III-311, 1983. (Abstract)

78. Krejci J, Granrud G, Salerno D, Griesbach D, Hodges M: Indecainide, a dose ranging efficacy trial for treatment of ventricular ectopic depolarizations. Circulation III-311, 1983. (Abstract)

79. Morganroth J, Horowitz LN, Mulhern C, Snyder S, Segal B: Indecainide: A new class IC antiarrhythmic: Pharmacokinetics, efficacy and safety. Circulation III-31, 1983. (Abstract)

80. Rae AP, Sokoloff NM, Webb CR, Spielman SR, Greenspan AN, Horowitz LN: Circulation Research, 1984. (in press)

8

ROLE OF BETA-BLOCKING AGENTS IN THE TREATMENT OF VENTRICULAR ARRHYTHMIAS

JOEL MORGANROTH, M.D.

INTRODUCTION

It is only in recent years that factors defining the efficacy of antiarrhythmic drugs have been clearly defined. In patients with chronic potentially lethal ventricular arrhythmias without hemodynamically significant symptoms[1] the most appropriate endpoint of drug efficacy should be the prevention of sudden cardiac death. The pharmacological endpoint of significant statistical reduction in ventricular ectopic frequency, however, has been the most widely applied "efficacy definition" to the study of new antiarrhythmic agents.

Antiarrhythmic drug efficacy when assessed in individual patients must take into consideration the high degree of spontaneous variability that exists in chronic ventricular ectopic frequency.[2] This high rate of spontaneous variability demands that proper statistical methodologies be used to evaluate drug testing protocols and that proper attention is given to trial design. Various statistical approaches have shown that an appropriate definition for drug efficacy (when comparing ambulatory ECG or Holter monitoring data obtained on drug therapy to baseline placebo control data) is a minimum of a 75% reduction in mean hourly ectopic frequency. This level of reduction insures that the observed decrease was due to the therapy and not to a consequence of spontaneous variability. Data analysis in such drug testing studies should be presented in terms of the percent of individual patients responding with this level of reduction rather than the pooling of ectopy data on placebo and drug therapy. This later practice may mask proper evaluation of efficacy and safety data such as the frequency of proarrhythmic events.

ROLE OF BETA BLOCKING AGENTS IN THE TREATMENT OF VENTRICULAR ARRHYTHMIAS

Since their availabilty in clinical medicine, beta-blocking agents have not been considered generally effective in the treatment of ventricular arrhythmias. Electrophysiologic studies demonstrated that beta blockers will block the catecholamine effect on Phase IV depolarization in automatic cells. Thus beta-blockers were found clinically useful as antiarrhythmic in patients whose ventricular ectopy occurred in settings such as the use of volatile anesthesia and exercise induced ventricular ectopy. Specific diseases as hypertrophic cardiomyopathy and mitral valve prolapse in which beta blocking agents have a primary role in their general therapy are also examples where beta blockers have had a role. Clinical cardiologists have, however, often employed beta blockers as adjunctive therapy with the traditional Class I antiarrhythmic drugs with the hope of improving ventricular arrhythmic therapy.

Table 1 details the clinical trials that have in recent years defined the efficacy of beta-blocking agents in the suppression of ventricular arrhythmias in which modern statistical approaches and trial designs were employed. These studies have used either homogeneous populations in which patients had predominantly coronary artery disease or mitral valve prolapse, or more commonly have employed populations with mixed cardiac conditions including cases with no structural heart disease. Efficacy criteria have ranged from 70-75% reduction in ectopic frequency and all these trials have used an adequate dose of the beta-blocking agent which beta-adrenergic blockade should occur.[3-11] The average percentage of patients responding to beta-adrenergic drug therapy in these trials is approximately 50-60%. Side effect profiles of the beta-adrenergic blocking agents are quite low and usually are less than 5%. This is in marked contrast to the average of 20-30% side effects expected with the use of Class I antiarrhythmic agents.[1] Irrespective of the extent or nature of cardioselectivity, presence of intrinsic sympathomimetic activity or membrane effect of beta-adrenergic blocking agents demonstrated an approximate 50% efficacy rate in patients tested suggesting that none of

TABLE 1

ORAL BETA-BLOCKING AGENTS IN CHRONIC VENTRICULAR ECTOPY (VPCS)
EFFICACY USING HOLTER MONITORING

Reference	Diagnosis	Drug	Efficacy in % of Pts Studied	% Reduction of VPCs for Efficacy	Dose Used in mg/day
Winkle[3]	Mixed	Propranolol	2/16 (50%)	>70%	240
Winkle[4]	MVP*	Propranolol	5/9 (56%)	>75%	40-320
Pratt[5]	CAD**	Metoprolol	7/13 (54%)	>70%	100-200
Woosley[6]	Mixed	Propranolol	24/32 (75%)	>70%	up to 960
DeSoyza[7]	Mixed	Acebutalol	11/20 (55%)	>70%	300
Koppes[8]	AMI***	Propranolol	18/32 (56%)	>70%	80-480
Schlemann[9]	Mixed	Nadolol	7/13 (54%)	>75%	80-640
Morganroth[10]	Mixed	Atenolol	7/12 (58%)	>70%	50-200
Podrid[11]	Mixed	Pindolol	21/42 (50%)	>50-90%	20-80

*Mitral Valve Prolapse
**Coronary Artery Disease
***Acute Myocardial Infarction

these various properties alter the antiarrhythmic potency of this class. Thus, the antiarrhythmic effect on ventricular arrhythmia frequency may be considered a class effect.

Patients with lethal ventricular arrhythmia in general, have severe underlying disturbance in cardiac function, and studies conducted in this group using only beta blocking agents have demonstrated poor responses, although adjunctive use of beta blockers with other classes of antiarrhythmic drugs, has antidotally improved efficacy. The potential of sotalol, a Class III agent with beta-blocking properties and marked effects on repolarization, has been reported to be highly efficacious in patients with recurrent sustained ventricular tachyarrhythmias and may, therefore, prove to be a most useful drug in this setting.[12]

BETA ADRENERGIC BLOCKING AGENTS IN THE PREVENTION OF SUDDEN CARDIAC DEATH

Recently, three major cooperative clinical trials had been conducted that demonstrate that beta-adrenergic blocking agents used in patients immediately after acute myocardial infarction, have demonstrated effectiveness in reducing sudden cardiac death.[13-15] Unfortunately, little data are available evaluating the antiarrhythmic effectiveness of the beta-blockers (propranolol, timolol, and metoprolol) in patients in these trials.

In the metoprolol trial[14] patients with acute myocardial infarction were treated with 15 mg of the drug intravenously on admission to the coronary care unit and then dosed for 3 months at 100 mg bid of metoprolol. This was a randomized, double-blind, parallel study in which half of the patients received metoprolol and the other half received matching placebo. Six hundred and ninety-eight patients were treated with metoprolol and 697 with placebo. Baseline comparisions between the two groups demonstrated comprability in many clinical and objective findings. The antiarrhythmic data[16] show there were 17 out of 697 episodes of ventricular fibrillation in the placebo group compared to only 6 out of 698 in those patients on metoprolol ($p<0.01$) and the metoprolol group demonstrated a marked decrease in the requirement for lidocaine therapy for ventricular ectopy in the coronary care unit (16 out of 698) compared to the placebo group (38 out of 697) ($p<.01$).

Preliminary information has been also available on the effect of propranolol on arrhythmias in the beta-blocker heart attack trial (BHAT).[17] This was a 31 center placebo controlled, double-blind randomized trial involving 3,837 post myocardial infarction patients. Propranolol was prescribed to 1,916 and placebo to 1,921. These agents were administered after baseline evaluation which occurred from the 5th to 21st day post myocardial infarction. After randomization in this period follow-up averaged 30 months. This trial demonstrated an approximate 25% decrease in mortality in those patients randomized to propranolol compared to placebo. Approximately 50% of the deaths in this trial (as is true in most trials of this nature) are due to sudden cardiac death suggesting that an arrhythmia may potentially be the primary mechanism causing this reduction. Holter monitoring was performed in patients at baseline and in a randomized subset of approximately 25% of the population at six weeks. Sixteen hundred and forty-one out of 1,916 patients randomized to propranolol (85.6%) had successful 24 hour Holter monitoring data at baseline compared to 1,630/1,921 patients randomized to placebo (84.9%), a nonsignificant difference. At six weeks, 407 patients on placebo and 419 patients on propranolol underwent a second 24 hr Holter monitor.

Irrespective of whether patients were on antiarrhythmic agents or not (approximately 15% of patients were on antiarrhythmic agents in this trial), the frequency of ventricular arrhythmia classes at baseline were comparable in patients in either the propranolol or placebo group. When comparing 6 week to baseline Holter monitoring data in this study, it was clear that there was an increase of approximately threefold in the prevalence of ventricular arrhythmia in those patients on placebo (Table 2), whereas in those patients on propranolol, there was a marked blunting of this increase in ventricular arrhythmia frequency which suggests an antiarrhythmic effect for propranolol.

Preliminary data have demonstrated that patients in this trial in whom successful reduction of ventricular arrhythmias occurred had a 0% prevalence of sudden cardiac death compared to approximately 7-8% in those in whom ventricular arrhythmias were suppressed.[18] Since this trial was not designed to determine the mechanism of sudden cardiac death prevention, these data are only to be considered preliminary.

Beta-blocking agents have important effects on ventricular fibrillation threshold. The efficacy of beta-adrenergic blocking agent in sudden cardiac death prevention may be on a basis of interfering with microreentry and its induction of ventricular fibrillation. Class I membrane stabilizing antiarrhythmic agents may be more important in suppressing macroreentry which generates ventricular tachycardia.

It is quite evident from the available data, that the "Class II" beta-blockers agents should be considered antiarrhythmic agents. In patients with benign and potentially lethal ventricular arrhythmias, the frequency of suppression of PVC's by beta blockers is comparable to that of other

TABLE 2. HOLTER MONITORING DATA AT BASELINE AND AT SIX WEEKS IN PATIENTS IN THE BHAT TRIAL

	IN % PATIENTS WITH A VENTRICULAR ARRHYTHMIA CLASS	
	AT BASELINE	SIX WEEKS
Placebo Group	33/407 (8.1%)	104/407 (25.6%)
	Z Value = 0.5	Z Value = 3.95
Propranolol Group	30/419 (7.2%)	61/419 (14.6%)

more standard antiarrhythmic agents (e.g. Class I drugs), yet the tolerance level of the beta-blocker class is much higher.[1]

In patients subjected to long-term cooperative clinical trials with beta-blockers in the early post-myocardial infarction period a marked reduction in sudden cardiac death has been demonstrated. The mechanism by which beta-blockers effect this important reduction is unknown. Since sudden cardiac death is primarily due to ventricular tachyarrhythmias which degenerate into ventricular fibrillation, and since beta-adrenergic blocking agents have an important effect on ventricular arrhythmia frequency and ventricular fibrillation threshold one can hypothesize that the reduction in sudden cardiac death prevalence due to beta blocking agents (which are the only agents to date that have shown prevention of sudden cardiac death) results from their antiarrhythmic properties. Further data are needed to clarify this issue.

REFERENCES

1. Morganroth, J: Premature ventricular complexes: Diagnosis and indications for therapy. Clinical Cardiology, JAMA 5:673-676, 1984.

2. Morganroth J, Michelson EL, Horowitz LN, Josephson ME, Pearlman AS, Dunkamn WB: Limitations of routine long-term electrocadiographic monitoring to assess ventricular ectopic frequency. Circulation 58:408, 1978.

3. Winkle RA, Gradman AH, Fitzgerald JW, Bell PA: Antiarrhythmic drug effect assessed from ventricular antiarrhythmic reduction in the ambulatory ECG and treadmill test. Comparison of propranolol, procainamide, and quinidine. Am J Cardiol 42:473, 1978.

4. Winkle RA, Lpes MG, Goodman DJ, Fitzgerald JW, Schroeder JS, Harrison DC: Propranolol for patients with mitral valve prolapse. Am Heart J. 93:422, 1977.

5. Pratt CM, Matlack J, Carney S, Waggoneer AD, Wichemeyer WW, Martin F, Miller RR: Effect of metoprolol in suppression of ventricular ectopic beats. Circulation 62:III, 1980.

6. Woosley RL, Kornhause D, Smith R, Reele S, Higgins SB, Nies AS, Shand DG, Oates JA: Suppression of chronic ventricular arrhythmias with propranolol. Circulation 60:819, 1979.

7. DeSoyza N, Kane JJ, Murphy ML, Laddu AR, Doherty JE, Bissett JK: The long-term suppression of ventricular arrhythmias by oral acebutolol in patients with coronary artery disease. Am Heart J 100:631, 1980.

8. Koppes GM, Beckmann CH, Jones FG: Propranolol therapy for aventricular arrhythmias two months post myocardial infarction. Am J Cardiol 46:322, 1980.

9. Schlemann M, Morganroth J, Reid P: Beta-blockade for suppression of high grade ventricular arrhythmias. J Am Coll Cardiol 1:613, 1983.

10. Morganroth J: Short-term evaluation of atenolol in hospitalized patients with chronic ventricular arrhythmias. DRUGS 5:181 185, 1983.

11. Podrid PJ, Lown B: Pindolol for ventricular arrhythmias. Am Heart J 104:491-496, 1982.

12. Senges J, Lengfelder W, Jauernig R, Czygan E, Brachmann J, Rigos I, Kubler W: Comparative effects of Sotalol, Metoprolol and Quinidine on sustained ventricular tachycardia. Circulation 66:II-142, 1982.

13. The Norweigan Multicenter Study Group. Timolol-induced reduction in mortality and re-infarction in patients surviving acute myocardial infaraction. N engl J Med 304:801-7, 1981.

14. Hjalmarsosn A, Elmfeldt D, Herlitz J, et al: Effect on mortality of Metoprolol in acute myocardial infarction: A double-blind randomized trial. Lancet 2:823-7, 1981.

15. Beta-blocker Heart Attack Trial Research Group: A randomized trial of propranolol in patients with acute myocardial infarction. I. Mortality results. JAMA 247:1707-1882, 1982.

16. Ryden L, Ariniego R, Arnman K, Herlitz J, Hjalmarson A, Holmberg S, Ryes C, Smedgard P, Svedberg K, Vedin A, Waagsein F, Waldenstrom A, Wilhelmsson C, Wedel H, Yamamoto M: A double-bline trial of Metoprolol in acute myocardial infarction. N Eng J Med 308:614-618, 1983.

17. Lichstein E, Morganroth J, Harrist R, Hubble E for the BHAT Study Group. Effect of propranolol on ventricular arrhythmia. The Beta-blocker Heart Attack Trial Experience. Circulation 67:II 1-5, 1983.

18. Morganroth J, Lichstein E, Hubble E, Harrist R: Reduction of chronic ventricular arrhythmias and sudden cardiac death in patients post myocardial infarction. Clin Res 31:207A, 1983.

9

STATUS OF CLASS III ANTIARRHYTHMIC DRUGS: AMIODARONE, BRETYLIUM AND SOTALOL

Leonard N. Horowitz, M.D.

Although seemingly disparate, the drugs which have been grouped in the Vaughan Williams classification as Class III have several unique similarities. In addition to their pharmacologic property of prolonging myocardial repolarization they were all introduced for clinical use and study for indications unrelated to cardiac arrhythmias. Therefore, it is important to remember that their development was not guided by the intention to develop an ideal antiarrhythmic agent. As Dr. Vaughan Williams has said:

> "Those who are concerned in the search for new and better drugs naturally like to think that their investigations are guided by pure reason, and that a devotion to logic and scientific principles will lead them inexorably to their goal. However, it has to be admitted that on occasion chance plays a role, and that a drug originally developed with one therapeutic aim in mind turns out to have scored a bull's eye on a neighbouring target.(1)"

Although I do not think the presently available agents in this class have scored that elusive bull's eye, they have certainly struck the neighbouring target. It is hoped that with the review of these agents, research directed toward the development of agents with the Class III property of prolongation of myocardial repolarization will yield a more ideal agent for the treatment of cardiac arrhythmias.

AMIODARONE

Amiodarone is a benzofuran compound which was originally introduced in the early 1960's as an antianginal agent. The agent was modestly successful for the treatment of angina in patients with coronary artery disease. Subsequently, pioneering studies by Rosenbaum and coworkers (2-3) showed that amiodarone had efficacy as an antiarrhythmic for treatment of patients with both supraventricular and ventricular tachyarrhythmias. Amiodarone has now been extensively studied and its antiarrhythmic properties have been defined.

The most distinctive direct electrophysiologic effect of amiodarone is the prolongation of repolarization (4). This property has been identified in virtually all cardiac tissue including sinus nodal, AV conduction system, atrial and ventricular fibers. Studies in intact animals and man have shown similar prolongation of refractoriness in atrial, AV conduction system and ventricular tissue (5). In addition, amiodarone produces noncompetitive sympathetic blockade and it antagonizes both alpha and beta adrenergic sympathetic responses (6).

Amiodarone has also been reported to reduce coronary and peripheral vascular resistance. When administered intravenously in experimental animals, amiodarone reduces systemic vascular resistance and arterial pressure without significantly affecting myocardial contractile force (7). Amiodarone produced approximately two-thirds of the coronary vasodilatory effect when compared to nitroglycerin. At high doses a negative inotropic effect can be produced. In patients with left ventricular dysfunction, amiodarone decreased coronary vascular resistance, mean arterial pressure, left ventricular end diastolic pressure and systemic vascular resistance and increased coronary sinus flow (8). Oral amiodarone produces variable but minor effects on left ventricular function as measured by radionuclide ventriculography (9). Although it is not certain, it appears that the minor negative inotropic effects are counter-balanced by vasodilatation and thus even in patients with left ventricular dysfunction clinically significant hemodynamic effects have not been seen frequently.

Amiodarone is available in both oral and intravenous preparations. The pharmacokinetics of amiodarone are quite complex. The oral preparation has a low bioavailability ranging from 30 to 50%. Its volume of distribution although difficult to measure is large and has been estimated to exceed 5,000 liters. It has a reduced clearance rate and the mechanisms of clearance are poorly understood. It has an extremely long half-life ranging in most patients from 25 to 50 days. There is an extremely wide range of half-lives ranging from 15 to 115 days. The elimination pattern also frequently shows a rebound in plasma concentrations 1 to 2 weeks after discontinuing therapy. This pattern has been thought to be due to rapid clearance from well perfused tissues with slow elimination from a poorly profused deep compartment (10).

Because of the unusual pharmacokinetics of this agent, therapy has generally been initiated with a loading dose. Some have suggested that because a three or four compartment pharmacokinetic model best describes amiodarone's pharmacokinetics loading should occur in three or four steps.(11) In clinical use, however, loading is generally achieved with a high initial dose of 1 to 2 grams per day for 1 to 2 weeks followed by maintenance doses.

Because the pharmocokinetics of amiodarone are poorly understood and changes in total body and plasma concentrations may occur over time, some have suggested that plasma concentrations of amiodarone should be monitored (9,12). Although this is controversial most recent evidence suggests that monitoring of amiodarone plasma concentrations may assist in control of the drug's toxicity and although less likely may be useful in identifying ineffective regimens.

Efficacy in Atrial Fibrillation

Amiodarone has been used both for the conversion of atrial fibrillation to sinus rhythm and for the maintenance of sinus rhythm in patients with recurrent paroxysmal atrial fibrillation. Successful conversion of atrial fibrillation to sinus rhythm with oral amiodarone has been reported in approximately 80% of patients (13). Maintenance of sinus rhythm for periods longer than a year can be expected in more than half of these patients. Although neither controlled nor double blinded this preliminary study suggests that amiodarone may have a role in converting atrial fibrillation to sinus rhythm.

Significantly more experience has been accumulated with the use of amiodarone for suppression of recurrent episodes of paroxysmal atrial fibrillation (Table 1). Rosenbaum initially reported complete suppression of this arrhythmia in 29 of 30 patients (3). Others have reported somewhat less impressive results; however, in these latter studies amiodarone has generally been used in patients who have been shown to be refractory to conventional therapy (10,12,14). There is little doubt from both early and more recent studies that amiodarone can effectively maintain sinus rhythm or at least make recurrent episodes of atrial fibrillation asymptomatic in a substantial proportion of patients who have not responded to other regimens.

Table 1. Efficacy of amiodarone in atrial fibrillation

Authors	N	Results		
		Excellent	Good	Poor
Rosenbaum et al (3)	30	29 (97%)	0	1 (3%)
Podrid and Lown (14)	20	16 (80%)	0	4 (20%)
Haffajee et al (12)	48	41 (85%)	0	9 (15%)
Zipes et al (10)	95	74 (78%)	6(6%)	15 (16%)

Amiodarone has been evaluated for the treatment of atrial fibrillation in the setting of the tachy-brady syndrome (15). Since amiodarone can suppress sinus node function (16) some patients with this syndrome may be worsened by the drug despite the fact that their atrial tachyarrhythmia is suppressed. Thus amiodarone therapy in conjunction with permanent pacemaking may be necessary in selected patients.

Efficacy in Paroxysmal Supraventricular Tachycardia

Experience has been somewhat more limited in the treatment of paroxysmal supraventricular tachycardia than in the therapy of atrial fibrillation. Rosenbaum has reported excellent suppression in 57 of 59 patients with paroxysmal supraventricular tachycardia (3). Subsequent studies, typically in patients who have previously been judged refractory to conventional therapy, showed somewhat less impressive results (10,12,14). Amiodarone has been reported to be particularly efficacious in patients with paroxysmal supraventricular tachycardia associated with the WPW syndrome (3,10,14). Suppression or significant amelioration of the tachycardia has been reported in 80 to 100% of patients. Of particular note is the fact that patients with the WPW syndrome may have both atrial fibrillation and paroxysmal supraventricular tachycardia and amiodarone may be particularly effective in such patients because of its ability to suppress both arrhythmias. Unlike most available antiarrhythmic agents, amiodarone prolongs repolarization in both the normal AV conduction system as well as accessory AV connections and thus is capable of having a beneficial effect in both types of supraventricular tachyarrhythmia.

Efficacy in Ventricular Tachyarrhythmias

Amiodarone has been reported to be effective in the treatment of patients with sustained life-threatening ventricular tachyarrhythmias (3,10,12,17-20). Initial studies suggested that amiodarone produced an excellent response and complete elimination of arrhythmia in greater than 80% of patients. However, with larger studies and longer follow-up periods it is now clear the drug is effective in 50 to 70 percent of patients and that a significant incidence of side effects which cause discontinuation can be expected (Table 2). The explanation for this wide range in response rates is not clear; however, significant differences in patient populations, methods of evaluation, loading protocols and duration of follow-up exist between studies and these may explain at least in part the discrepancies.

Table 2. Efficacy of amiodarone in ventricular tachyarrhythmias

Authors	N	Results	
		Good	Poor
Haffajee et al (12)	50	38 (76%)	12 (24%)
Nademanee et al (17)	96	70 (73%)	26 (27%)
Morady et al (13)	23	21 (91%)	2 (9%)
Kaski et al (19)	23	15 (65%)	8 (35%)
Zipes et al (10)	196	129 (66%)	67 (34%)
Horowitz et al (20)	100	62 (62%)	38 (38%)

Like many antiarrhythmic drugs amiodarone is capable of worsening ventricular arrhythmia in some patients. The incidence with which this occurs with amiodarone ranges from

5 to 10% (21). This is comparable to most other released and investigational antiarrhythmic agents (22). Amiodarone, in fact, may produce this effect less often than the typically used antiarrhythmic agents.

Some controversy has developed regarding the utility of invasive electrophysiologic testing for the evaluation of amiodarone therapy. Initially, it was suggested that the clinical course in patients with recurrent ventricular tachyarrhythmias treated with amiodarone could not be predicted by clinical electrophysiologic studies as they could be when these arrhythmias were treated with our other agents. More recently several studies have shown that this may not be the case (20). In patients in whom arrhythmias were no longer inducible following amiodarone treatment the clinical course was excellent. In patients in whom the arrhythmia remained inducible a significant recurrence rate was noted. Furthermore, the severity of the recurrence can generally be predicted by the electrophysiologic study performed after amiodarone loading and thus, clinically useful information can be obtained that may alter the patient's subsequent management.

Adverse Effects of Amiodarone

It is generally agreed that amiodarone has a relatively high efficacy in most supraventricular and ventricular tachyarrhythmias. The incidence of serious adverse effects, however, has only recently been recognized. Many of these side effects may be mild and relatively unimportant, however, others may be life-threatening.

Amiodarone can produce two types of skin reactions (21,23). Photosensitivity which may range from an increased propensity to sunburn to intense burning, erythema and swelling are exposed areas is common but rarely is severe enough to cause discontinuation of the drug. Frequently the severity of the reaction can be ameliorated by reduction in dose. A rare side effect of amiodarone is the slate gray or purple pigmentation of the face (21,23). This is frequently seen in patients who have been taking high doses for prolonged periods of time. It is associated with lipofuscin deposits in the skin and elastic degeneration in the skin.

The major opthalmologic side effect of amiodarone is corneal microdeposits (24). These are bilateral, symmetrical brownish linear opacities generally found initially along the corneal meniscus. They are detected by slit-lamp examination and generally progress in severity during therapy. Although uncommonly associated with symptoms, a small number of patients (5%) may report photophobia, colored halos or blurred vision. Microdeposits will regress with doseage reduction or discontinuation but this may take years.

Gastrointestinal side effects are generally noted early in therapy principally during loading. These are nausea, loss of appetite or constipation. They resolve during maintenance therapy and since they are generally self-limiting, uncommonly cause discontinuation of therapy.

During chronic therapy liver function abnormalities have been reported. These are not uncommon, being seen in between 10 and 20% of patients. Evidence of clinical hepatic dysfunction, however, is uncommon (23). Liver function abnormalities are not commonly a cause of drug discontinuation.

Neuromuscular toxicity from amiodarone is generally uncommon, however, it can take any of several forms (23). Amiodarone has been reported to produce a proximal myopathy which effects primarily the lower extremities. This may be severe and cause incapacitation. Although this is uncommon, it typically requires discontinuation. The myopathy resolves several months after drug discontinuation. Amiodarone has also been reported to produce a sensorimotor neuropathy, which has a typical glove-and-stocking distribution. Nerve conduction studies have shown reduced velocity. This neuropathy is uncommon and resolves with discontinuation. Others have reported cerebella ataxia and tremor which are generally mild but may be severe and cause discontinuation of therapy.

Thyroid abnormalities have been prominantly reported during therapy with amiodarone (21,23). Although many patients will have mild elevations of the T4 following initiation of amiodarone therapy, clinical hypo-and hyperthyroidism are uncommon. They are thought to be related to the large amount of organic iodine which is present in amiodarone and are generally treated in the usual fashion. This abnormality does not generally resolve with discontinuation of amiodarone.

One of the most significant toxicities of amiodarone is fibrosing alveolitis. This may be difficult to differentiate from congestive heart failure or other pulmonary syndromes which patients with cardiac disease develop. This reaction can produce dyspnea, nonproductive cough, rales or severe pulmonary insufficiency. Symptoms are generally associated with interstitial infiltrates on the chest x-ray but on occasion may present as less obvious pulmonary abnormalities sometimes resembling a neoplasm. It is uncommon having been reported in between 1 and 10% of patients and generally requires drug discontinuation (21,23).

Although amiodarone has adrenergic blocking properties, decreases of myocardial function have not been prominantly reported among its toxicities. There is a small incidence of worsening of congestive heart failure following the initiation of amiodarone therapy, however, this is probably related to bradycardia produced by the drug rather than a direct inotropic effect. Amiodarone has significant effects on the cardiac conduction system and occasionally intraventricular conduction disturbances or sinus node dysfunction may be precipitated by the drug and require pacemaker implantation. This is not a common toxicity, however, it must be considered in a patient who develops worsening symptoms and bradycardia while on the drug.

Aggravation of ventricular arrhythmias by amiodarone has been reported but is not common (21).

Drug interactions are a significant concern with amiodarone and may lead to toxicity (10,26). Amiodarone has an intense interaction with warfarin and potentiates its anticoagulant effect. The maintenance dose of warfarin must be reduced sometimes by as much as half when amiodarone is begun. Amiodarone increases serum digoxin concentrations promptly following initiation of therapy. Amiodarone interacts with other antiarrhythmic agents and significant toxicity may occur if the dosage of adjunctive antiarrhythmic agents is not carefully adjusted.

In summary, amiodarone is a potent antiarrhythmic agent with demonstrated efficacy against a broad spectrum of arrhythmias. It also has significant toxicity and careful monitoring is necessary for appropriate therapy with this agent.

SOTALOL

Sotalol is a beta adrenergic blocking agent which was originally introduced for the treatment of hypertension and angina pectoris. It is a competitive beta blocker with approximately one half the potency of propranolol (27). In addition to its beta blocking activity it also prolongs myocardial repolarization. This class III activity of sotalol was initially described in rabbit atrial and ventricular tissue in vitro and subsequently confirmed in man (28,29).

Sotalol is available in both intravenous and oral preparations. The oral preparation has a high bioavailability (greater than 95%) and a moderately long half life ranging from 15 to 20 hours (30).

The clinical electrophysiologic effects of sotalol have been studied and found to be precisely what would be expected in a beta blocking agent with the ability to delay myocardial repolarization (31,32). Sinus cycle length significantly increased. The QT and JT intervals were increased. AV nodal function was depressed as evidenced by increases in the AH interval and effective and functional refractory periods of the AV node. No change in the HV interval was noted. The atrial ventricular and accessory pathway effective refractory periods were prolonged. Edvardsson and coworkers have also shown a significant increase in monophasic action potential in the ventricle following the intravenous injection of sotalol (33).

Electrophysiologic Evaluation of Sotalol for Supraventricular and Ventricular Arrhythmias

Sotalol has been evaluated during electrophysiologic testing both in patients with supraventricular and ventricular tachycardia. Nathan and coworkers showed that sotalol given during atrioventricular reentrant supraventricular tachycardia produced termination in 5 of 12 patients (31). Sotalol has also been compared to metoprolol using electrophysiologic testing in patients with paroxysmal supraventricular tachycardia. Sotalol prevented induction of sustained PSVT in 10 of 17 patients (59%) whereas metoprolol prevented induction in 4 of 17 (28%) ($p < 0.05$) (34). In both studies, the site of action of sotalol has been identified as the antegrade pathway in the AV node and the retrograde pathway if this was an accessory pathway. In those patients in whom sotalol failed to prevent induction of the SVT the tachycardia cycle length was significantly increased. Standard beta blockers have not been shown to have an effect on extra-nodal accessory pathways as has been demonstrated for sotalol. In patients with sustained ventricular tachycardia during electrophysiologic studies sotalol has been reported to prevent induction of the arrhythmia in 12 of 18 patients (67%) when it was administered intravenously (35). Prolongation of the QT interval and increases in ventricular refractoriness were observed, however, these changes did not predict the efficacy of the drug. In 10 of 11 patients in whom intravenous sotalol was effective in prevention of induction of sustained ventricular tachycardia, oral sotalol reproduced the same effect. Long-term prophylaxis in patients in whom sotalol prevented induction of tachycardia has been good.

Adverse Effects of Sotalol

Sotalol can produce any of the typical beta blocker side effects and these do not appear to be more marked with this agent than any other beta blocking drug. Sotalol can be expected to behave like other non-cardio-selective beta blockers which possesses no intrinsic sympathomimetic effects. Sotalol has also been reported to produce the syndrome of prolongation of the QT interval and torsades de pointes (36). The frequency of this adverse effect is variable but low. Like with other agents, this toxicity is probably most commonly seen in patients with severe myocardial disease and other proarrhythmic abnormalities such as electrolyte disturbance or ischemia.

Sotalol is an intriguing beta blocker with significant Class III antiarrhythmic efficacy. Its place in therapy and a more precise understanding of its toxicities are required before recommendations about its clinical use can be made with authority.

BRETYLIUM

Bretylium is a quaternary ammonium compound which is an adrenergic blocking agent. Introduced in the 1950's as an antihypertensive (37) it never gained wide popularity because of tachyphylaxis and significant undesirable side effects during chronic oral therapy. Approximately twenty years ago bretylium's antiarrhythmic activity was first reported (38) and bretylium tosylate was released for treatment of resistant ventricular arrhythmias in the late 1970's.

Bretylium is taken up by peripheral adrenergic nerve terminals. After an initial phase of norepinephrine release which may produce a sympathetic effect it produces inhibition of norepinephrine release. There is no evidence of central sympathetic blocking activity by bretylium.

In vitro studies of bretylium have shown an increase in the action potential duration and the effective refractory period measured in ventricular tissue (39, 40). Unlike most antiarrhythmic agents, however, bretylium does not depress automaticity or prolong the effective refractory period relative to the action potential duration. In intact animals, intravenous bretylium produces an increase in sinus cycle length, depression of atrioventricular conduction and significant increases in atrial and ventricular effective refractory periods (41).

Bretylium has been reported to increase myocardial contractility based upon its sympathetic releasing property (42, 43). In clinical situations, however, this effect may be exquisitely dose dependent and an increased inotropic response has not always been reported. In man, bretylium produces a two-phased hemodynamic response (44). Initially, heart rate and blood pressure increase and following this heart rate, arterial pressure and systemic vascular resistance fall. Cardiac index does not generally change significantly.

The pharmacokinetics of intravenous bretylium are not complex (37). Its half-life is 7-10 hours with considerable individual variation. Bretylium is is not significantly metabolized.

Efficacy of Bretylium in Ventricular Tachyarrhythmias

Bretylium has been reported to be effective in ventricular arrhythmias occurring in a variety of settings including myocardial infarction, cardiac surgery and cardiopulmonary resuscitation (37). In general, the arrhythmias have been drug resistant and recurrent and bretylium is typically used after other antiarrhythmic agents have been found to be ineffective.

Bretylium has been used as a "drug of last resort" in patients with recurrent ventricular arrhythmias. Clinical reports have shown an excellent efficacy in such patients (37,45-48). These studies are summarized in Table 3. Typical of these studies is the one by Bernstein and Koch-Weser in which 30 patients with refractory ventricular tachyarrhythmias were treated with bretylium (48). Following parenteral administration 18 of the 30 patients had no further major ventricular arrhythmias. As will be discussed subsequently, hypotension was the most frequent side effect and is sometimes therapy-limiting.

Table 3. Efficacy of bretylium in resistant VT/VF

Authors		N	Arrhythmia Results		
			Controlled	Not Controlled	Survival
Terry et al	(45)	10	7 (70%)	3 (30%)	5 (50%)
Dhurandhar et al	(46)	18	15 (83%)	3 (17%)	8 (44%)
Cohen et al	(47)	25	16 (64%)	9 (36%)	18 (72%)
Bernstein et al	(48)	30	23 (77%)	7 (23%)	15 (50%)

Bretylium has been evaluated in the setting of cardiopulmonary resuscitation to identify a beneficial effect in promoting resuscitation. Initially, bretylium was used in patients who proved resistant to resuscitation. In one typical study, bretylium was administered in patients in whom ventricular fibrillation was resistant and failed to respond to standard cardiopulmonary resuscitation with lidocaine and other antiarrhythmic drugs (49). In 20 of 27 patients, a stable rhythm was achieved following administration of bretylium and 12 of these 20 patients survived and were discharged from the hospital. More recently, bretylium has been evaluated in protocols in which it has been compared to placebo or lidocaine. In the placebo trial, 59 patients who presented with cardiopulmonary arrest were given either bretylium (10 mg/kg) or placebo by rapid intravenous bolus in addition to standard cardiopulmonary resuscitation techniques (50). Thirty-five percent of patients presenting with ventricular fibrillation or asystole who received bretylium were successfully resuscitated whereas 6% of patients who received placebo survived ($p < 0.05$). In a similar study comparing bretylium with lidocaine an organized rhythm was restored in 89% of patients receiving bretylium and 93% of patients receiving lidocaine and a stable profusing rhythm was achieved in 58% of bretylium patients and 60% of lidocaine patients. Hospital discharge was achieved in 34% of patients who had received bretylium and 26% of the patients who had received lidocaine. In this study there was no significant difference between lidocaine and bretylium.

Bretylium has received modest attention in electrophysiologic laboratories. In the study by Greene and coworkers bretylium failed to prevent induction of VT or VF in a group of 5 patients in whom it was tested.

Adverse Effects of Bretylium

The major undesirable side effects of parenterally administered bretylium is hypotension. Proximal hypotension occurs in virtually all patients and hypotension may occur even in patients who remain supine. Occasionally transient hypertension following bretylium administration may be a clinical problem.

Bretylium has been found to be an effective parenteral agent for the management of refractory ventricular arrhythmias. Like amiodarone and sotalol it possesses a class III effect and may, in certain circumstances, be the agent of choice. Its long-term use is limited by the unavailability of an acceptable oral form.

1. Vaughn Williams EM: The discovery of the antiarrhythmic action of amiodarone. In Simpson WT, Caldwell ADS (eds) Amiodarone in Cardiac Arrhythmias. International Congress and Symposium Series NO 16, Academic Press, London 1979, p.1

2. Rosenbaum MB, Chiale PA, Ryba D, Elizari MV: Control of tachyarrhythmias associated with Wolff-Parkinson-White syndrome by amiodarone hydrochloride. Am J Cardiol 34: 215-23, 1974

3. Rosenbaum MB, Chiale PA, Halpern MS, Nau GJ, Pryzbylski J, Levi RJ, Lazzari JO, Elizari MV: Clinical efficacy of amiodarone as an antiarrhythmic agent. Am J Cardiol 38:934-44, 1976

4. Singh BN, Vaughan Williams EM: Effects of amiodarone: A very antianginal drug on cardiac muscle. Br J Pharmacol 39:657-659, 1970

5. Saksena S, Rothbart ST, Cappello G: Chronic effects of amiodarone in patients with refractory ventricular tachycardia. Int J Cardiol 3:339-351, 1983

6. Charlier R, Delaunois G, Bauthier J, Deltour G: Pharmacology of amiodarone, an antianginal drug with a new biological profile. Arzneimittelforsch 18:1408-17, 1968

7. Singh BN: Amiodarone: Historical development and pharmacologic profile. Am Heart J 106:788-97, 1983

8. DeBoer LWV, Nosta JJ, Kloner RA, Braunwald E: Studies of amiodarone during experimental myocardial infarction: Beneficial effects on hemodynamics and infarct size. Circ. 65:508-12, 1982

9. Rotmensch HH, Belhassen B, Swanson BN, Shoshani D, Spielman SR, Greenspon AJ, Greenspan AM, Vlasses PH, Horowitz LN: Relationships between steady-state serum amiodarone concentrations, antiarrhythmic efficacy and toxicity. Annals Int Med 101:462-469, 1984

10. Zipes DP, Prystowsky EN, Heger JJ: Amiodarone: Electrophysiologic actions, pharmacokinetics and clinical effects. J Am Coll Cardiol 3:1059-71, 1984

11. Siddoway LA, McAllister CB, Wilkinson GR, Roden DM, Woosley RL: Amiodarone dosing: A proposal based upon its pharmacokinetics. Am Heart J 106:951-6, 1983

12. Haffajee CI, Love JC, Canada AT, Lesko LJ, Asdourian G, Alpert JS: Clinical pharmacokinetics and efficacy of amiodarone for refractory tachyarrhythmias. Circ 67:1347-1355, 1983

13. Santos AL, Alexio AM, Landeiro J, Luis AS: Conversion of atrial fibrillation to sinus rhythm with amiodarone. Acta Med Portuguesa 1:15-21, 1979

14. Podrid PJ, Lown B: Amiodarone therapy in symptomatic, sustained refractory atrial and ventricular tachyarrhythmias. Am Heart J 101:374-379, 1981

15. Brown AK, Primhak RA, Newton P: Use of amiodarone in bradycardia-tachycardia syndrome. Br Heart J 490:1149-1978.

16. Touboul P, Atallah G, Gressard A, Kirkorian G: Effects of amiodarone on sinus node in man. Br Heart J 42:573-1979

17. Nademanee K, Singh BN, Hendrickson J, Intrarachot V, Lopez B, Feld G, Cannom DS, Weiss JL: Amiodarone in refractory life-threatening ventricular arrhythmia. Ann Int Med 98:577-584, 1983

18. Morady F, Scheinman MM, Hess DS, Sung RJ, Shen E, Shapiro W: Electrophysiologic testing in the management of survivors of out-of-hospital cardiac arrest. Am J Cardiol 51:85-89, 1983

19. Kaski JC, Girotti LA, Messuti H, Rutitzky B, Rosenbaum MB: Long-term management of sustained recurrent symptomatic ventricular tachycardia with amiodarone. Circ 64:273-279, 1981

20. Horowitz LN, Greenspan AM, Spielman SR, Webb CR, Morganroth J, Rotmensch H, Sokoloff NM, Rae AP, Segal BL, Kay HR: Usefulness of Electrophysiologic Testing in Evaluation of Amiodarone Therapy for Sustained Ventricular Tachyarrhythmias Associated with Coronary Artery Disease, Amer J Cardiol (in press)

21. McGovern B, Garan H, Ruskin JN: Serious adverse effects of amiodarone. Clin. Cardiol 7:131-137, 1984

22. Velebit V, Podrid P, Lown B, Cohen BH, Graboys TB: Aggravation and provocation of ventricular arrhythmias by antiarrhythmic drugs. Arc 65:886-894, 1982

23. Harris L, McKenna WJ, Rowland E, Holt DW, Storey GCA, Krikler DM: Side effects of long-term amiodarone therapy. Arc 67:45-51, 1983.

24. Verin P, Gendre P, Barchewitz G, Laurent-Bronchat G, Yacoubi M, Morax S: Thesaurismose corneenne par amiodarone: donnees recentes. Arch Opthalmol 31:581-586, 1971

25. Marcus FI, Fontaine GH, Frank R, Grosgogeat Y: Clinical pharmacology and therapeutic applications of the antiarrhythmic agent, amiodarone. Am Heart J 101:480-493, 1981

26. Marcus FI: Drug interactions with amiodarone. Am Heart J 106: 924-929, 1983

27. Lish PM, Weikel JH, Dungan KW: Pharmacological and toxicological properties of 2 beta-adrenergic receptor antagonists. J. Pharmac Exp Ther 149:161-173, 1965

28. Singh BN, Vaughan Williams EM: A third class of antiarrhythmic action. Effects on atrial and ventricular intracellular potentials and other pharmacological actions on cardiac muscle of MJ1999 and AH 3474. Br. J Pharmacol 39:675-687, 1970

29. Bennett DH: Acute prolongation of myocardial refractoriness by sotalol. Br Heart J 47:521-526, 1982

30. Anttila M, Arstila M, Pfeffer M, Tikkaneu R, Vallinkoski V, Sunguist H: Human pharm toxicol 39:118-128, 1976

31. Nathan AW, Hellestrand KJ, Bexton RS, Ward DE, Spurrell RAJ, Camm AJ: Electrophysiological effects of sotalol - just another beta blocker? Br Heart J 47:515-520, 1982

32. Touboul P, Atallah G, Kirkorian G, Lamaud M, Moleur P: Clinical electrophysiology of intravenous sotalol, a beta-blocking drug with class III antiarrhythmic properties. Am Heart J 107:888-895, 1984

33. Edvardsson N, Hirsch I, Emanuelsson H, Ponten J, Olsson SB: Sotalol-induced delayed ventricular repolarization in man. Eur Heart J 1:335-343, 1980

34. Rizos I, Senges J, Jauernig R, Lengfelder W, Czygan E, Brachmann J, Kubler W: Differential effects of sotalol and metoprolol on induction of paroxysmal supraventricular tachycardia. Am J Cardiol 53:1022-1027, 1984

35. Senges J, Lengfelder W, Jauernig R, Czygan E, Brachmann J, Rizos I, Cobbe S, Kubler W: Electrophysiologic testing in assessment of therapy with sotalol for sustained ventricular tachycardia. Circ. 69:577-584, 1984

36. Neuvonen PJ, Elonen E: Sotalol, prolonged Q-Tc interval and tachyarrhythmias. Lancet 1:106, 1982

37. Heissenbuttel RH, Bigger JT: Bretylium tosylate: A newly available antiarrhythmic drug for ventricular arrhythmias. Ann Int Med 91:229-238, 1979

38. Leveque PE: Antiarrhythmic action of bretylium. Nature 207:203-204, 1965

39. Bigger JT, Jr, Jaffee CC: The effect of bretylium tosylate on the electrophysiologic properties of ventricular muscle and purkinje fibers. Am J cardiol 27:82-92, 1971

40. Wit AL, Steiner C, Damato AN. Electrophysiologic effects of bretylium tosylate on single fibers of the canine specialized conducting system and ventricle. J Pharmacol Exp Ther. 173: 344-356,1970

41. Waxman MB, Wallace AG. Electrophysiologic effects of bretylium tosylate on the heart. J Pharmacol Exp Ther. 183:264-274, 1972

42. Saunders WH, Behar VS. The effect of bretylium tosylate on ventricular function. Am J Med Sci 265:287-295, 1973

43. Markis JE, Koch-Weser J. Characteristics and mechanism of inotroic and chronotropic actions of bretylium tosylate. J Pharmacol Exp Ther. 178:94-102, 1971

44. Chatterjee K, Mandel WJ, Vyden JK, Parmley WW, Forrester JS. Cardiovascular effects of bretylium tosylate in acute myocardial infarction. JAMA 223:757-60, 1973

45. Terry G, Vellani CW, Higgins MR, Doig A. Bretylium tosylate in treatment of refractory ventricular arrhythmias complicating myocardial infarction. Br Heart J 32:21-5, 1970

46. Dhuradhar RW, Teasdale SJ, Mahon WA. Bretylium tosylate in the management of refractory ventricular fibrillation. Can Med Assoc J 105:161-5, 1971

47. Cohen HC, Gozo EG Jr, Langendorf R, Kaplan BM, Chan A, Pick A, Click G: Response of resistant ventricular tachycardia to bretylium relation to site of ectopic focus and location of myocardial disease. Circulation 47:331-40, 1973

48. Bernstein JG, Koch-Weser J. Effectiveness of bretylium tosylate against refractory ventricular arrhythmias. Circulation 45:1024-34, 1972

49. Holder DA, Sniderman AD, Fraser G, Fallen EL. Experience with bretylium tosylate by a hospital cardiac arrest team. Circulation 55:541-4, 1977

50. Nowak RM, Bodnar TJ, Dronen S, Gentzkow G, Tomlanovich MC: Bretylium tosylate as initial treatment for cardiopulmonary arrest. Randomized comparison with placebo. Ann Emerg Med 10:404-407, 1981

51. Haynes RE, Chinn TL, Copass MK, Cobb LA: Comparison of bretylium tosylate and lidocaine in management of out-of-hospital ventricular fibrillation: a randomized clinical trial. Am J Cardiol 48:353-356, 1981

52. Greene HL, Werner JA, Gross BW, Sears GK: Failure of bretylium to suppress inducible ventricular tachycardia. Am Heart J 105:717-721, 1983

PANEL ON STATUS OF SPECIFIC ANTIARRHYTHMIC AGENTS

Dr. Morganroth: Should patients have antiarrhythmic agents prescribed as outpatients or should such drugs be used only when monitoring patients in hospital? Is this decision dependent on the drug selected or does it depend more on the clinical setting in which the arrhythmia occurs? What should we advise physicians or what should be in labeling relative to how to handle the initiation of antiarrhythmic drug therapy?

Dr. Pratt: We are often defensive about the investigational drug. If we take the ones we know more about, (e.g. quinidine) I think there is enough data now to say that there is a relatively safe group and a relatively dangerous group of patients. Those patients that are on digoxin and have left ventricular dysfunction and have QT prolongation are ones that should have drug administered while being monitored. I think as we develop more experience with the investigational agents, we will also be able to detail a patient profile at risk. In the design of the cardiac arrhythmia pilot study by the NIH all of the agents are given as in patients. I think it is more our lack of knowledge than it is that we absolutely know that hospitalization needs to be done. Many of us have investigated agents in out-patient studies in stable patients with good ventricular function and frequent ventricular arrhythmias and never have a really serious incident of proarrhythmia.

Dr. Morganroth: What is your recommendation for the patient who has ventricular arrhythmias one month after a myocardial infarct and in the office and you detect PVC's on the ECG and the patient is asymptomatic. A Holter monitors 8 hours of frequent runs of non-sustained VT and an average of 48 PVC's per hour. Do you admit the patient to the hospital to institute antiarrhythmic agents, offer no treatment, or treat by a prescription as an outpatient.

Dr. Pratt: I think that there are certain drugs that we feel fairly safe in beginning as out-patient therapy. The beta-blockers would be one group. The type I agents in that patient have some risk.

Dr. Horowitz: I think it is the arrhythmia in the clinical setting rather than a specific drug. I think I know the available agents well, but I would not start a patient who had just been resuscitated from a cardiac arrest on quinidine as an outpatient. On the other hand, I would feel perfectly comfortable starting a patient with no heart disease and paroxsymal atrial fibrillation on quinidine as an outpatient. I think the patient with the benign arrhythmia, if you are forced to treat them by circumstances, that they probably can be treated with most agents as an outpatient. Patients with life-threatening arrhythmias probably should have their therapy initiated in hospital no matter what therapy you are choosing and the patients with potentially life-threatening arrhythmias are more controversial. If a patient has a recent myocardial infarction, and marked left ventricular dysfunction, I think no matter what you start them on, probably they ought to be in hospital. On the other hand if the patient has frequent VPC's and maybe a triplet on the Holter monitor and either mild or no left ventricular dysfunction, I would feel comfortable starting them as out patients.

Dr. Fogoros: I would like to comment briefly on a side effect of amiodarone which has received very little attention from the literature but which I feel is of more than academic interest now that many of us are using amiodarone in conjunction with the automatic implantable defibrillator. That side effect is creating refractoriness to cardioversion. In a series of 76 patients with sustained ventricular tachycardia treated with amiodarone at Stanford, 5 patients were seen who had previously had several episodes of successful cardioversion. All 5 died with their first episode of recurrent ventricular tachycardia after being placed on amiodarone when that tachycardia could no longer be successfully cardioverted. We commented on these 5 patients in our description of 76 patients which appeared in Circulation last year and at the time stated that we couldn't be sure that amiodarone caused refractoriness to cardioversuion. Since that time, I have seen an additional patient who fairly convincingly had refractoriness to defibrillation created by amiodarone and which was reported in the Annals of Internal Medicine in May of this year. Briefly, this gentlemen had about 4 episodes of successful cardioversion prior to being placed on amiodarone. No drugs were found during serial drug testing and he was placed on amiodarone because there was going to be a two month delay in getting an automatic defibrillator for him. At the time of his defibrillator implant, he had then been on amiodarone for two months. We induced ventricular fibrillation after placing the defibrillating leads on the heart and could not defibrillate him at all via the leads. We had to shock him 3 times with 400 wattseconds in order to defibrillate him trans-thoracically. We stopped the amiodarone and 2½ months later took him back to the OR and repeated defibrillation thresholds and his defibrillation thresholds after being off amiodarone for 2½ months were less than 10 wattseconds. I wonder if any of the panelists have seen this side effect or if any of the other members of the audience have seen this.

Dr. Lipicky: I have not seen any reports to that effect

Dr. Reder: I would just like to inquire if a recommendation is forthcoming that many of these agents like amiodarone be started in patients while in hospital particularly when they are still under investigational status, who is going to pay for the hospitalizations with the advent of the DRG's?

Dr. Morganroth: I think that like any decision in clinical medicine you have to weigh the risks and benefits and now to some extent the finances. If we had to admit to a monitored facility everybody that was to start on an antiarrhythmic drug, I think that the medical care system would be financially strained. The experience of many clinical trials with instituting quinidine and more potent drugs as flecainide of encainide has been one of low rates asymptomatic proarrhythmic deaths and in fact none in the flecanide-quinidine experience in 280 people. I agree with Dr. Horowitz's view point that whether drugs need to be started in hospital depends on the clinical group being studied.

Dr. Lin: I wonder how one can do a placebo in cardiopulmonary rescuscitation. Can one use normal saline versus bretylium in such studies.

Dr. Horowitz: In on the street rescusitation by paramedics in many cities, it is the protocol to shock the patient before doing anything, including starting an IV and in this study, the administration of saline was tantamount to shocking the patient and not giving any medical treatment and then the active treatment arm was administration of bretylium. I think when those studies were done, there was no definitive evidence that it was absolutely mandatory to administer an agent in this setting and I would bet that most paramedic teams to this day do not administer either bretylium or lidocaine before they give the first cardioversion.

Dr. Lin: What about the diurnal variations of arrhythmias? In other words, how do you report the data from the Holter monitor which might be affected by the diurnal variations of arrhythmias.

Dr. Morganroth: Diurnal variation is an important issue to some people but not to others. Some patients at rest or sleep will drop their PVC frequency but many will also increase the frequency and many stay the same. I don't want to say it is 1/3, 1/3, 1/3 because it has not been that well studied and the population size has been small. We have not found it necessary to break Holter data into these intervals.

Dr. Pratt: A lot of investigators have tried to correct arrhythmia for heart rate. Obviously heart rate is higher during the day, if you remember Dr. Morganroth's point about the beta blocker trial, that the major effect of propranolol was to blunt the day time increase in arrhythmias, that has an awful lot to do with the heart rate effect. I think that this might be one of the major determinants.

Dr. Altman: I have a question for Dr. Lipicky. Since there is quite a wide variability in class III agents, what would you recommend in terms of an optimal agent for comparison for new drugs in this category?

Dr. Lipicky: I guess placebo. Does that answer your question? I didn't mean to be facetious.

Dr. Temple: We have to ask what comparative agent for what purpose? Is it a sudden death trial or a VPB trial. There are lots of things you could use in the latter. It is hard to know what to use in the former. The conversation about what to do when you start on a drug has identified only two alternatives, one being nothing and the other being hospitalization. There obviously are some things you could do if it were deemed worth while. Short of that for the relatively short half life drugs with no metabolite, you could keep the patient around while the peak blood levels were reached and look at his QT interval. Maybe that would take 2 hours and he would have to sit around. You could send him out with a Holter in place and get the result of that 24 hours later. These are some things that you could do. I am not saying they are things you should do, but they are cheaper than hospitalization and I guess the question I would put is when you do send someone out on an antiarrhythmic drug when do you see him again and how soon do you do the next Holter? If there is going to be some monitoring that you think is right, why not do it rather early?

Dr. Horowitz: The reason to hospitalize the patient for initiation of therapy is that you are afraid that they are going to die. If someone is going to get nausea and vomiting, you don't need them in the hospital to detect that. Unfortunately, that is a rather black or white end point and the problem that I have with saying we'll see you daily or ask you to transmit an electrocardiogram to us daily by using some trans-telphonic device or will ask you to call our research nurse and tell us how you are doing is that really is a momentary check and if the patient happens to die on the second day they would have been O.K. on the first day. It also implies that there is always going to be some evidence of warning in either an exam or test or whatever. I am not convinced looking at all of the agents that there is always a warning for proarrhythmia that you can't say well if your Q-T goes from x to y you are in the dangerous zone or if your QRS duration goes from x to y that you are in the dangerous zone. I think you have to look at a group. I don't think that you can pick out individual patients. I think that if you look at groups of patients with different types of arrhythmias, you can say the odds of an proarrhythmic event in this group are so small that it doesn't justify the cost of hospitalizing them to evaluate. If there is a 1 in a 1000 chance of something bad happening, then I don't think that we would hospitalize that patient. If there is a 1 in 10 chance, I think we would, knowing that 9 of the 10 patients are going to sit there smiling in the hospital. I am not sure that monitoring a patient for a couple of hours until they reach peak levels, even if it is a drug that doesn't have metabolites.

Dr. Temple: You get most of the way to your final dose with a short half life drug after the first dose. Not all the way, but most of the way.

Dr. Horowitz: I think if the drug doesn't have complicated metabolism and doesn't have complicated kinetics, that you can see proarrhythmia in several days or a week which tends to work out well in scheduling and do your testing, your blood level measurements and Holters. If it is complicated metabolism, you might have to see them in a week, but you would do your testing sometime later.

Dr. Temple: What do you look for at a week when you see them. What are you looking for that is going to shape your decision.

Dr. Horowitz: To see if they have any toxicity, to see their electrocardiogram. I don't generally look for efficacy, or I am not critically concerned about efficacy at 1 week in a patient who is now an outpatient, because if they had a life-threatening arrhythmia I was satisfied when they went home that they at least had more than a modicum of protection. So at one week I am not terribly concerned and if all they have is PVC's I would rather make sure I am in a steady state situation.

Dr. Temple: There isn't for example for some of the drugs some Q-T interval beyond which you would say I don't want to risk this anymore.

Dr. Horowitz: Let's pick a number that everybody in the audience would. If I saw someone come back with a Q-T of .65 I would say O.K. without pulling the rip cord. I wouldn't have the patient rush back every day for an electrocardiogram because they have a Q-T of .51 I wouldn't know what to do with it.

Dr. Temple: You wouldn't change your therapy?

Dr. Horowitz: I wouldn't know what to do.

Dr. Temple: Then there isn't any point in monitoring that.

Dr. Lipicky: Yes, I simply want to be sure that there really was no predicted value for the interval and indeed if what you say if it is very very long, that then you know what to do but in between you don't, then what monitoring would mean is monitoring all the time. A snapshot wouldn't help.

Dr. Horowitz: Correct. Even if we knew that we would have to monitor all the time and I am not sure we even know the limits that we would use to cut off if we could monitor all the time.

Dr. Morganroth: The sicker the patient and the more severe the left ventricular dysfunction and the more severe the ventricular arrhythmia is, the more likely that proarrhythmia may occur and the more likely for death. That is the group that you want to keep in the hospital. The ones that are more towards no heart disease and a few PVC's the risk is low and you could be very safe.

Dr. Borland: Are we being mislead on the estimates of a proarrhythmic affect because these patients are the ones with higher incidence, have a compromised left ventricular function and in many instances in these long term trials, when there appears to be a proarrhythmic affect, patients are withdrawn from that therapy and then changed to another antiarrhythmic rather than being taken off all antiarrhythmics with baseline Holter recording obtained?

Dr. Morganroth: I think the question is a good one and that is how do we know when there is a proarrhythmic event on a long term trial that it is really a proarrhythmic event or a clinical event that is part of their natural history of the arrhythmia. Shouldn't we take them off the drug, put them on a period of no drug and the redose to see if you can re-establish the proarrhythmia.

Dr. Pratt: We have had over 200 patients as outpatients on ethmozine over the past 5 years at Baylor and I would say that about 1/4 of these were started in hospital. The incidence of proarrhythmia with that drug in our hands is very low. That is maybe, 2 to 3%, maybe 4 to 5 patients. There was frequently a change in clinical status for one reason or another or even without change in clinical status, once we took them off the drug and did do the monitoring that you suggested, their arrhythmia baseline now was new and what was considered baseline before was now classified as proarrhythmia based on the change. This led us to a research project that we have now submitted for publication in which we looked at the natural history of arrhythmia. What one does in that situation is to get an initial baseline of 3 to 5 days of monitoring to have have someone in a drug trial at 1 to 2 years later, stop the drug and give them another placebo period in blinded fashion and see how the baseline compares to 1 to 2 years. At least 25% of patients will have at least 50% lower frequency and in about 20% of patients it is at least 100% higher. In some patients hands that is proarrhythmia so I think we need to remind ourselves that the natural history of arrhythmia isn't that once you get a frequency you can use that for life, and if one is going to trigger proarrhythmia to any drug, they need to re-establish baseline at the time they've done that and make sure clinical status, especially left ventricular function, angina, etc. hasn't changed.

Dr. Morganroth: My own recommendation is to reintroduce placebo at least at 6 months, because a year seems so long to me. Patients have such a high likelihood of having their condition change or new drugs added if one waits a year.

Dr. Ruffy: I like to get back to the issue of hospitalization versus outpatient treatment. I am personally impressed by the number of patients who have been referred to me for an episode of sudden cardiac death which occurred within 24 to 48 hours after initiation of antiarrhythmic treatment, and in whose only long term successful management was to withdraw all antiarrhythmic therapy. These include patients with and without structural heart disease. I don't think we had an argument regarding the management of patients who have had spontaneous episodes of life-threatening arrhythmias off of antiarrhythmic treatments. I don't think we have an argument regarding our absence of knowledge regarding the need of antiarrhythmic or the benefits of antiarrhythmic treatment in patients who had not had symptomatic episodes. I think Dr. Morganroth just made that point clearly a while ago. Therefore, if we decide to treat a patient who has not been symptomatic, just in case, not knowing what we are doing, I think the least we owe this patient is to maximize his/her safety and I think we must hospitalize these patients for a period of time which will allow at least 70% of them to be detected in case they have a fatal episode.

Dr. Temple: You don't know how to do that. If you take people who have had one episode of sudden death and don't treat them at all, they don't die immediately the next day. Over the next year they have maybe a 50% chance of doing that. To take just a nother example. If you make someone susceptible to sudden death by putting them on quinidine, he doesn't get that tomorrow necessarily. He gets it sometime, so it is a really hard to imaging the kind of study you do except by putting him through programmed electrical stimulation which would be an interesting study to do sometime in sequential people put on quinidine.

Dr. Temple: The idea of a proarrhythmic effect is like any other adverse effect, you are only really sure you have it if you can reproduce it with a re-challenge. In the person who as far as I know has the most data for all drugs on this sort of thing is Lown and Podrid. What they say is that when they say something was proarrhythmic, that means that they have reproduced whatever they have defined as the proarrhythmic event, more than once, that is if it is 100% increase in VPB rate they have done it at least twice. That is what they said so that those figures may reflect something that is meaningful if in fact that is what they do. But certainly if you just see somebody go up a little bit, that is not very impressive. Why isn't it fairly easy to re-challenge somebody in a controlled setting if you think that a drug has had a proarrhythmic event.

Dr. Morganroth: I think that the answer to that is if the proarrhythmia is a four-fold increase in PVC frequency, there should be no reason not to. If the proarrhythmic event is harder to resuscitate or sustain Vt, I wouldn't think it's ethical to do that so I think it depends on the type of proarrhythmic event.

Dr. Temple: Where the person was very sick, being treated for an emergent arrhythmia, it is obviously very difficult to determine whether the drug has done it or whether it just occurred. I mean it is only when you see it in somebody who had no serious arrhythmia and then suddenly develops V-tach that you would be suspicious.

Dr. Somberg: Dr. Temple, the Lown group reproduced only the baseline. They didn't reproduce the intervention in their published study on VPB frequency worsening. I would say, why reproduce the arrhythmia on the drug? Why not just count it against the drug and go on. I mean we all study drugs, but our basic business, and I must admit I am in collaboration with Dr. Ruffy on this point, that our basic point is to benefit the patient. I would rather count it against the drug and go on to something else then to try to reproduce things in that and I will be very interested in trying to figure out how to use PES to look at arrhythmogenicity, but one thing I have really excluded is to give the drug because I just don't want to have a potential fatality on my hands and I am more interested in testing drugs 2,3,4,5,6 and just strike one against that drug.

Dr. Nacarelli: With the Podrid data, I am not sure that any of the investigators in the room are in agreement that disopyramide is less arrhythmogenic than tocainide, mexilitene or propranolol, however, given that tocainide, mexilitene, flecainide, amiodarone may be available commercially soon, I would be interested in the panel, given the patient that we commonly see, get away from the VT/VF patient, the patient with 38% ejection fraction, maybe six monthes post infarction, 15,000 PVC's, 1,000 couplets, 200 3-6 beat runs of VT, somebody I think most of us would treat. You see that patient in the office. Giving some of the newer drugs we have available to us, some of which may have less proarrhythmias, are there any favorites that they would use in the outpatient settings. Is there a threshold that they didn't work, that they would admit him to the hospital.

Dr. Morganroth: I don't think we have the data to answer the question in terms of comparative drug responses in populations. We do have some comparative trials of drug x versus y.

Dr. Somberg: I think the proarrhythmic effect is not idiosyncratic. It is probably due to the inherent electrophysiology of the agent, but I think it has to be like quinidine or procainamide, it may be protective in a given patient, while one may work and the other may not. It is also true that one may be proarrhythmic and the other not. In the electrophysiology laboratory, that certainly can be the case and I have anecdotally, and I know Dr. Horowitz has probably seen more of these type of cases, but someone who may be inducible on a class I agent may be protected by a class I agent and that is possible. I would say you can't use the classification which is very helpful to keep drugs straight and to maybe consider potential interactions, but in the opposition, I would say to the classification system, it doesn't say here is a patient who all IC agents would be proarrhythmic on. It is also related to the dose of the drug and their metabolism as well.

Dr. Chun: The problem really is the sudden death. That is the real problem we have. I have asked all the drug companies who make antiarrhythmic drugs to give us the data of the control studies and also emergency program studies. Give us the treatment drug and how many days the patient was on the drug before the patient died. We shall have all the data summarized very soon.

Dr. Morganroth: The data may be difficult to understand. Let's say you get data on 6-8 drugs and you see a spread of time from institution of therapy to sudden death. We don't know if the event is proarrhythmic or just part of the natural history. There is no way to tell which. I think that if the death occurs in about a week it might be proarrhythmic and if the death occurs after 2-3 months it is probably not proarrhythmic.

Dr. Chun: If the deaths occur real early after starting drug, than the chance of that being drug induced death is a lot more than if the death occurred 3 or 4 months later.

Dr. Lipicky: It depends on why you are treatingand who the patient is. I mean the temporal relationship doesn't seem to me to add or subtract to the degree of confidence that one would have that that event occurred because of the drug.

Dr. Somberg: It depends on the prior natural history also. I mean if they were dying every day previously and they go on a drug and they die 4 days later, you can't say that is proarrhythmic. If they have never had a sudden death episode and they die three days later, that may be more suggestive.

Dr. Morganroth: So again, it depends on whether you are dealing with potentially lethal or lethal patients.

Dr. Greenspan: There may be one other wild card in this. I'm not sure you can really make the assumption that sudden deaths which occur early are the ones that are due to the proarrhythmic event and those that occur later are not because there are many interactions between the drug and other metabolic conditions that can then produce a proarrhythmic event and these of course can occur at any time in the natural history of the patient's therapy. I would caution against using a break point and being so confident about it. It may be helpful as a starting point.

Dr. Temple: I just want to object slightly to what I think is a sort of know-nothing attitude towards counting something as proarrhythmic just because you have seen it. The discussion we just had said that before you treated something to drugs like a sudden death, you have to think about who is being treated and whether there is another explanation. The consequence of merely counting every event as though it is really drug induced, is that companies that are willing to allow their drug to be used in relatively sick people will find that their drug is denounced. It will have all kinds of things attributed to it and they won't want to do that anymore, and then you will have drug studied in completely well people prior to being put out into the marketplace where they will be given to sick people and you will never know what happens. You do have to think about these things it seems to me and just saying I'll count it against the drug and that way I will protect my patients best, I don't think answers the question. I'm not saying you should expose people to high lethal possibilities but you do want to think about the event is really the drug or not.

III. SPECIAL CONSIDERATIONS AND FDA STANDARDS

10

SOME THOUGHTS ON EFFICACY TRIALS OF ANTIARRHYTHMIC AGENTS: PITFALLS OF MIMICKING CLINICAL PRACTICE

Gordon Pledger, Ph.D.

It is understandable that in designing clinical drug trials researchers would have a desire to "imitate life", i.e. to use the test drug in the trial in much the same way as it might be used in medical practice. A design and analysis which parallel the expected clinical use of the drug might be seen as lessening the problems of extrapolation which so often arise after a clinical trial is completed. However, this desire to mimic clinical practice should be severely tempered when on considers the differing goals involved. The goal in clinical practice is successful treatment of an individual patient, and although the clinician might like to know how much of the patient's improvement was due to the therapy used, that information is of secondary importance and is generally unobtainable. In a clinical trial the goal is to characterize the intervention in some way e.g. to show that a test drug has antiarrhythmic activity. This goal necessitates a strong emphasis on experimental controls and reduction of bias, an emphasis which dictates experimental conditions generally unlike those of clinical practice. A clear illustration of one way in which mimicking clinical practice can cause a clinical trial to get the wrong answer was given by Temple (1). He presented examples of antihypertensive trials in which doses were titrated to achieve a specified goal blood pressure. This method of varying dose, common in clinical practice, gave a misleading picture of the dose response curve. Even if a study is well-designed, a statistical analysis which departs from the trial design in an effort to demonstrate the "true drug effect" can defeat the purpose of the study. An example of this is the analysis of only so called "evaluable" patients, i.e. patients who comply with the protocol in certain ways which may be specified either in advance or post-hoc. It is now generally recognized that such analyses can introduce serious biases. Based on what I have seen in the medical literature as well as submissions to FDA certain pitfalls of mimicking clinical practice are relevant to antiarrhythmic drug trials at this time, and I would like briefly to discuss some of these.

Much of what I have to say has been said before, but I think it bears repeating in the context of antiarrhythmic drug trials. As evidence of this I would cite a recent article by Berry and Fox (2) which discusses statistical analysis of the data from an individual patient as opposed to analysis of groups of patients. They "propose the application of linear regression to PVC histories to ascertain drug effect in individual patients". Berry and Fox favor individual over grouped analysis because of the desire to know which patients respond to the drug and what the magnitude of the effect is. In their words: "... analysis by grouping can hide drug effectiveness, but (even) when such an analysis concludes that a drug is effective, it is difficult to say how effective...". What they say is true. Grouping <u>can</u> obscure effectiveness when many of the patients studied are not responsive to the test drug, and the group estimate of magnitude of effect in a controlled study can not usually be taken as directly applicable to clinical practice. However, these problems have more to do with the objective and the design of the study than with the statistical analysis. A randomized, controlled, clinical trial attempts to impose experimental conditions under which observed differences between the test and control groups can be attributed to treatment. (There must, of course, be an allowance for chance occurrences, but dependence on unproven assumptions and models, either biological or statistical, should be minimized.) The experimental conditions resulting from this objective are highly artificial. They may include double-blinding, fixed treatment dosage, restrictions on concomitant therapies, and limiting participation to specially chosen patients and physicians. There is no basis for expecting estimates constructed under such restrictions, using either group of individual data, to be representative of future "real world" experience. This unavailability of estimates relevant to clinical practice is due not to shortcomings in the statistical analysis but to the carefully controlled conditions generally necessary in order to scientifically demonstrate activity of a therapeutic agent. We should keep in mind the words of Tukey (3):

> "It is right for each physician to <u>want</u> to know about the behavior to be expected from an intervention or therapy when applied to his particular individual patient... It is not right however for a physician to <u>expect</u> to know this..."

an initial period of single-blind treatment with a marketed antiarrhythmic drug A. After this initial period only those patients who do not have a "satisfactory" reduction in PVC's are randomized into the double-blind portion of the study, which might compare the marketed drug A to a test drug B. It should be noted that the motivation for this type of design, rather than to mimic clinical practice, could be either to demonstrate that the new drug works for patients who are refractory to the marketed drug or, as suggested by Temple (4), to "enrich" the study population and thereby obtain a more sensitive test of the efficacy of the new drug B. Whatever the motivation for choosing this sort of design, there needs to be a careful consideration of potential bias introduced due to pre-screening the patient population with one of the study treatments. For instance, in the example I have given one group of patients received the treatment sequence AA while the other group received AB. How should one interpret a result showing a greater mean PVC reduction in the AB group? Did screening with drug A successfully identify patients who are non-responsive to that treatment? What are the effects of withdrawing drug A? Is there more unblinding in the AB group than in the AA group? Could treatment A be inferior to placebo in the patients studied? Screening with drug A has introduced an additional complexity into the trial design, and questions such as these must be answered, either explicitly or implicitly, in order to interpret the study results. It is clearly preferable, in the interest of scientific credibility and persuasiveness, to address these questions explicitly.

At this time it is unclear what real costs are associated with these designs which decide to randomize a patient or not based on observations made while the patient is receiving one of the drugs to be compared. However, two things are clear: (i) these designs have at least the potential to mislead, and (ii) they are being used with increasing frequency. For these reasons the possible problems in interpretation of such studies deserve serious thought and discussion in the clinical trial community.

REFERENCES

1. Temple, R.: Evaluating dose-response relationships for new drugs. Presented at the annual meeting of the Society for Clinical Trials, 1983.
2. Berry DA and Fox TL: Regression analysis applied to PVC histories: a statistical procedure for evaluating antiarrhythmic drug efficacy. Statistics in Medicine 2:331-343, 1983.
3. Tukey JW: Some thoughts on clinical trials, especially problems of multiplicity. Science 98:679-684, 1977.
4. Temple, R. Issues in study design: case studies from the FDA. Presented at the annual meeting of the Society for Clinical Trials, 1984.

11

USE OF A COMPUTER IN THE NEW DRUG EVALUATION PROCESS

STEWART J. EHRREICH, Ph.D.

1. INTRODUCTION

The clinical reviewer for New Drug Applications (NDA) or Investigational New Drug Exemptions (IND's) is faced with the sheer weight of the applications and NDA submissions have gone well above 1,000 volumes in size. Since each volume is about 500 pages long, this is about one half million pages to review.

The evaluation process can be simply considered as a critique of the application to determine if the new drug is both safe and efficacious. The clinical studies are now supplied in hard copy form and the reviewer is forced to either accept analyses provided by the sponsor, methodically do his own, or ask the sponsor for special analyses. In most cases all three options are exercised.

A computer generated data base and proper software will allow the reviewer instant access to patient records and make it possible to do various analyses and display data in many different ways. The purpose of this presentation is to discuss the advantages, disadvantages and potential problems when the new drug evaluation process is automated. This is based on some limited experience in our own division but does not represent the final official determination by the Agency on the usefulness of this procedure.

2. WHY COMPUTERIZE?

There are three possible reasons to computerize the review process:

The computer allows more efficient (faster, more accurate) review of the application.

The amount of paper supplied by the sponsor is reduced.

The reviewer can find data and do correlations which would be difficult or impossible to do using the present manual methods.

No experiment has yet been performed which would allow us to determine if any of the above is indeed true, nor has our recent experience been long enough to make possible any final decisions.

3. ACCESS TO CLINICAL DATA - WHAT'S NEEDED (SEE FIG.1)

The medical reviewers should have:

3.1. Horizontal Access. Access to all data by patient initials and/or study number in a friendly, menu-driven, forms-oriented system.

3.2. Vertical Access. Access to all variables across patients (by study arm or drug group) in an easy-to-use, english-language type of database retrieval system; e.g., print PVC's/hour, PR intervals, ADVERSE REACTIONS OF ALL PATIENTS, sorted by drug and time on therapy.

3.3. Summary Tables. (supplied by sponsor) Similar to currently available formats.

3.4. Capability of generating new tables/summaries in VIRTUALLY ANY FORMAT ON-LINE BY REVIEW OFFICER (AD HOC OR AT CONCLUSION OF STUDY).

4. ADVANTAGES/DISADVANTAGES OF COMPUTER ACCESS OF DATA BY THE REVIEWER.

4.1. Introduction.

4.1.1. Computer generated information allows the reviewer to spot check specific data on a patient, or review all data including analysis of cause of death. The reviewer can tabulate data by grouping variables different from those used in sponsor generated summaries. (e.g. PVC's by drug and sex) and in addition the reviewer can run statistical analyses in subsets using different statistical procedures (See fig 2).

4.2. Advantages.

Horizontal review (fig. 1) is easier than leafing through CRF's (clinical record forms) and eliminates the need to call the sponsor to clarify illegible data. It also eliminates inconvenience, cost of storing and accessing boxes of paper. Present space at the Food and Drug Administration is at a critically low level.

Figure 1

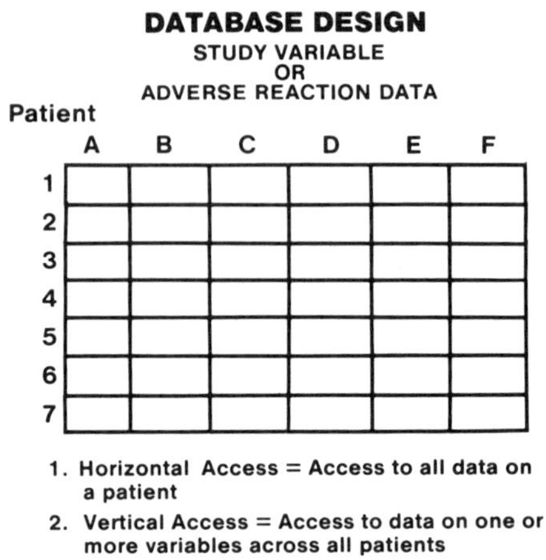

Figure 2

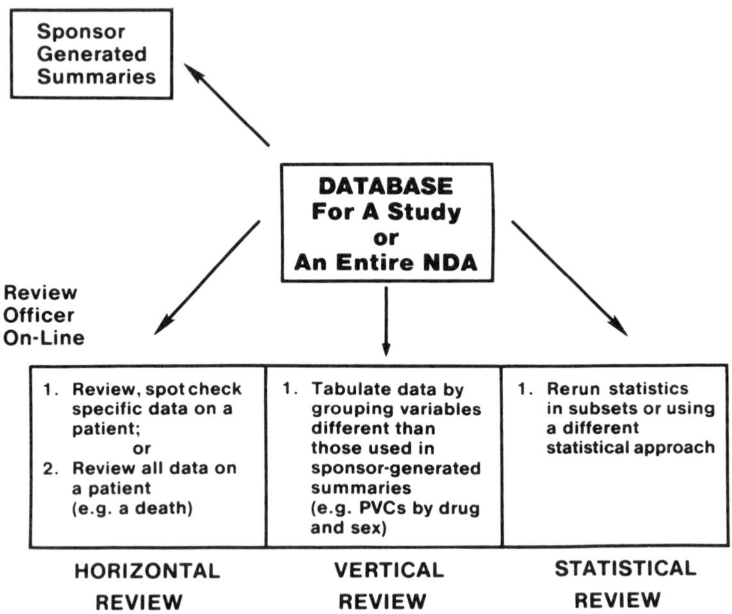

Vertical review allows the clinical reviewer to look at data in a different way that that proposed by the sponsor. Tabular review capability exists when questions are fresh in the reviewer's mind, perhaps making for a more rapid review.

4.3. Logistical problems.

4.3.1. The initial dollar cost may be high (especially to the sponsor) if the data must be in computer-accessible form. All investigators must be supplied with a terminal by the sponsor and personnel must be fully trained in its proper use. Breakdowns, and failure by personnel to use the system properly must be dealt with quickly and efficiently. An emergency phone number or other provision should be made available. The FDA reviewers' system must be compatable with the sponsor's, and as "user-friendly" as possible.

Studies of small numbers of patients (one or two dozen) may not be cost-effective to put in computer format, but no determination as to what is too small has been made yet.

Medical reviewers and investigators and their personnel may suffer from "computer phobia" and be reluctant to learn the techniques. The system must be designed to be handled by nervous or uncertain people and must have simple procedures to make quick changes if data is input incorrectly.

4.3.2. The investigators must also learn to be comfortable with automation and be able to understand how data can be input directly into the terminal provided.

4.4. On-line vs off-line approaches.

The on-line analysis of clinical data as it is generated (in the IND phase) brings up interesting issues whose solution is not clear.. Most important of these is the ability of the reviewer to determine the current status of a clinical trial before it is completed. Thus, the reviewer can make subgroup analyses, correlations, determinations and draw conclusions while a study is underway, so it is possible that FDA may determine that a clinical study should be terminated prematurely. The other side of the coin however is that patients may be spared from continuation in a study which would not ordinarily be stopped for sometime to come, or ethical considerations may force a decision to discontinue patients on placebo or active drug. While larger multi-center studies utilize Data Monitoring Committees and other safety measures, the use of the on-line data retrieval capability may provide significant data to review before the regularly scheduled meetings of such committees and force decisions at earlier stages in the study. This may or may not be considered a disadvantage to the patient but premature or incorrect conclusions could destroy a valuable trial.

5. LEGAL ASPECTS.

What is the legality, responsibilities and ethical considerations in on-line review? In the event of automation, does the reviewer now have a responsibility which he did not have before? Consider for a moment that the reviewer is able to instantly determine that an adverse reaction has occurred, provided that he goes to his terminal and logs into the system. What if he doesnt go to the terminal for several days and misses an important event? Under present (manual, hard copy) procedures the reviewer might not have been aware for days, weeks, or even months of such events. But automation changes these circumstances!

The advent of the computer however, does not remove the responsibility of the sponsor to make the Agency aware of adverse drug reactions (ADR's) and automation only raises questions relative to how this can now be done. Our preliminary evaluation indicates that reviewers using automation would be no more liable for failure to uncover data than they are at present.

The sponsor must simply ensure that all data is collected and is placed in a format which is totally retrievable by the reviewer. This involves careful planning and discussions between sponsor, investigator, reviewers, and systems personnel from both the Agency and the clinical trial officials. Everyone must be comfortable with the system and all questions must be worked out before the process begins.

At the present time it is not possible for the Agency to accept computer generated 1639 forms in lieu of hard copy. It is conceivable however, with a properly functioning system and complete reviewer awareness, that a substitution may be made by having all such forms generated on printers at the Agency. It is not certain at present how many such printers should be available, or whether a central location will suffice.

This is still in an investigative status but it is safe to say that automation will never remove obligations of sponsors to promptly make data available to reviewers and the Agency will not give up its rights to make a determination that a trial should be modified or even terminated.

6. COST VERSUS BENEFIT CONSIDERATIONS. WHAT WILL BE GAINED OR LOST BY AUTOMATION?

6.1. Saving to sponsor.

6.1.1. The sponsor's CRA (clinical research associate) may save a certain amount of travel and inspection time by being able to monitor a clinical study in the same way as the FDA reviewer. Errors or violations in the protocol may be discovered quickly and adjustments made by telephone or electronic mail to investigators. In this way patients may be salvageable in a certain protocol if corrective measures are quickly taken. This can severely reduce the cost of lost patients.

6.2. Savings to investigator.

With proper training, data need only be input once, and is retrievable quickly for review for publications etc.

Emergency information is quickly available and can be flagged and studies can be made easier to evaluate.

7. Audit of information.

By "going computer" what will be the method which will be used to evaluate the credibility of a study? At present, investigations into ongoing or completed clinical trials involve painstaking "detective work" by Agency investigators and the sponsor's CRA's.

The computer must not permit violations to occur in protocol design, compliance, false or modified data entries, and the like. Use of the system should occur only if the following procedures are observed:

7.1. Audit trials should allow complete documentation of who initially made an entry, when and what changes were made, and by whom.

7.2. The investigator at the study site should not enter data himself but it should be entered by a nurse or technician. (This decreases the error rate by an estimated 50%.)

7.3. The Sponsor should not be able to "write" on the system (enter data or make changes) but only receive data.

7.4. The FDA must be able to obtain all data which a steering or safety committee would normally be able to access. This would mean that the Agency must know what study medication was used (by symbol A, B, C, etc) even if the protocol totally blinds both investigator and safety or steering committees.
While this would not divulge the medications used to the Agency it makes it possible to determine if patterns exist on a particular study medication as far as safety (or efficacy) data by analysing results of patients with the same coded symbol treatment.

Should a serious adverse reaction be identified by the Agency reviewers using the coded procedure, such information could be brought to the attention of the appropriate data monitoring committee and appropriate action taken.

7.5. Computer fraud (hidden "keys" to data, illegal access to locked files, failure of write-protection and the like) opens up new issues of security and maintenance of secrecy. These issues have hardly been addressed.

8. Summary.

The Agency is currently using computer generated data on an experimental basis to determine the value of this method of review of applications. At the present time it is not possible to judge the value of automation and it may well be that such procedures are not worth persuing. On the other hand there is a great awareness on the part of the Agency and the sponsors of investigational drugs that a shortened more efficient review process with smaller volumes of paper would benefit everyone.

The results of the experiment are not yet ready for interpretation.

12

SUDDEN CARDIAC DEATH - FAILURE OR EFFECT OF ANTIARRHYTHMIC DRUG THERAPY?

RONALD W.F. CAMPBELL, M.D.

INTRODUCTION

Cardiac arrhythmias probably underlie the majority of incidents of sudden cardiac death. Data from out-of-hospital rescue squads[1,2], from coronary care units[3] and from chance events captured on dynamic ECG recorders[4] show that ventricular tachyarrhythmias (ventricular tachycardia, ventricular flutter and ventricular fibrillation) are the commonest arrhythmias implicated.

Reducing sudden cardiac death is an important clinical goal which, although elusive, appears technically feasible. The problems lie in identifying individuals at particular risk and in applying an effective modifying management. Despite a plethora of antiarrhythmic drugs with scientifically proven effectiveness against ventricular arrhythmias, little has been achieved in reducing sudden cardiac death. Of many possible explanations for this, the following are important:

a) Therapy is given to the wrong individuals.
b) Therapy is ineffective.
c) Therapy is effective but adverse effects encourage non-compliance.
d) Therapy is effective for some individuals but kills others.

SUDDEN DEATH RISK PREDICTION

Sudden cardiac death is a recognised complication of the acute phase of myocardial infarction, the congenital long QT syndromes, aortic stenosis, cardiomyopathies, chronic ischaemic heart disease, the pre-excitation syndromes, mitral valve prolapse and has an unexpected but important incidence in apparently normal individuals. Not all patients with these diagnoses share the same risk and despite much effort expended in refining individual risk prediction, risk assessment remains clinically imprecise. The highest risk situations identified are for patients in the first hours of acute myocardial infarction[5], for patients with bundle branch block and axis shift complicating anteroseptal myocardial infarction[6,7], for survivors of secondary ventricular fibrillation complicating acute myocardial infarction (ventricular fibrillation in the presence of shock and/or failure)[8], for survivors of non-infarction out-of-hospital ventricular fibrillation[1] and for patients with ventricular tachyarrhythmias complicating cardiomyopathies[9].

In none of these examples is the degree of risk higher than 20-30% per event (in the case of myocardial infarction) or per year. The implications of such imprecision is that management strategies will be applied to a number of patients who are not destined to succumb to sudden cardiac death. Therefore, any positive effects of applied therapy may be obscured by administration to patients who can derive no benefit.

EFFICACY OF ANTIARRHYTHMIC THERAPY

Broad criteria for drug efficacy are well established but differ according to the application. Efficacy, as perceived by drug regulatory authorities may be very different from that sought by the physician when treating an individual patient.

Efficacy - clinical problems.

The arrhythmia. Determining the effect of administered therapy depends upon knowledge of the index arrhythmia prior to drug administration but the natural history of arrhythmias in man is not well defined. The unstable nature of life threatening ventricular arrhythmias and their devastating consequences for the patient should they occur out-of-hospital, has hampered direct examination of antiarrhythmic drug actions.

Prophylaxis. If a drug is to be used for prophylaxis then no direct assessment of drug effect is possible.

Non-compliance. Drug effects unrelated to antiarrhythmic activity are relevant as even modest unwanted effects may encourage non-compliance with consequent fluctuating or sub-therapuetic plasma levels. Scientifically conducted studies do not reliably assess this problem as enrolled patients, knowing the importance of their participation, may feel an obligation to continue dosing.

Efficacy - ventricular ectopic beats.

A standard technique for determining antiarrhythmic drug efficacy is the supression of ventricular ectopic beats in patients who show a chronic stable pattern. The often remarkable frequency of these phenomena permits complex statistical examination of the effect of administered therapy but the results may be applicable to only that highly selected patient population. There is no support for the hypothesis that a drug effective against ventricular ectopic beats will necessarily be effective in the prevention of life threatening ventricular tachyarrhythmias.

Efficacy - invasive electrophysiology.

Invasive electrophysiological testing has revealed much about fundamental drug action but has proved disappointing in predicting individual patient risk and has had little role in evaluating drug efficacy against ventricular fibrillation. However, it reflects an important advance for evaluating and treating other life-threatening ventricular tachyarrhythmias as it provides direct evidence of drug efficacy[10]. However, the technique is applicable to only those arrhythmias which can be reliably induced by programmed stimulation. By implication the underlying arrhythmias are of re-entrant or triggered automatic type, the former based on a permanent arhythmia substrate. Invasive electrophysiology cannot reliably address drug evaluation for life-threatening ventricular tachyarrhythmias which depend upon a substrate created and modified by ischaemia, autonomic influences or circulating catacholamines.

Efficacy - evidence of sudden death reduction.

To date, no Vaughan Williams Class I drug (quinidine,

procainamide, disopyramide, mexiletine, tocainide, flecainide, encainide, lorcainide, propafenone) or Class IV drug (verapamil, diltiazem) has, in a controlled study, been proven to reduce the incidence of sudden cardiac death[11]. Nonetheless, numerous case reports and personal experience suggest that in selected individual patients, these agents can reduce mortality[12,13,14]. By contrast, there is impressive evidence that beta-adrenoreceptor blocking drugs reduce sudden cardiac death. The relatively modest effects of these agents on isolated ventricular ectopic beats led to speculation that their action was not directly antiarrhythmic but this now seems unlikely. Recent reports confirm that beta-adrenoreceptor blockers can prevent ventricular fibrillation[15] - an arrhythmia electrophysiologically different from ventricular ectopic beats. Reduction of sudden cardiac death by beta-blockers has been reported in survivors of acute infarction[16,17,18], and in patients with the congenital long QT syndromes[19]. Amiodarone, a Vaughan Williams class III drug, may reduce the incidence of sudden cardiac death when administered to patients with tachyarrhythmias complicating hypertrophic cardiomyopathy[20] but further follow-up is required to confirm this result.

ARRHYTHMOGENICITY

Definition.

Arrhythmogenicity is the property of an antiarrhythmic agent to aggravate an existing arrhythmia or to produce a new arrhythmia but its clinical identification is difficult. Most suggested criteria are empirical. Important demonstrations of drug arrhythmogenicity have occurred in invasive electrophysiological tests when the following responses have been observed:

a) Induction of a new tachycarrhythmia
b) Acceleration of the native tachyarrhythmia.
c) Degeneration of the native tachyarrhythmia to ventricular fibrillation, the native arrhythmia having previously been stable.

d) Conversion of the spontaneously terminating native to one which is not self-terminating.

The relevance of these responses observed in electrophysiological testing to clinical usage is unclear but not surprisingly, few patients have been given long term oral therapy with an agent which, in invasive testing, aggravated the native event.

Although not truly arrhythmogenic, deterioration of the haemodynamic status of the patient with aggravated effects of the electrically unchanged native arrhythmia may result from antiarrhythmic drug use and underlie some instances of arrhythmia degeneration.

Evidence from published studies.

The literature abounds with case reports of drug arrhythmogenicity which usefully alert us to this complication but which give no information of the risk of this event. Identification of arrhythmogenicity in prospective studies has been relatively rare but has been noted with encainide[21,22], lidoflazine[23], quinidine[24,25], amiodarone[26] and flecainide[27]. In these reports, recognition of the occurrence depended upon electrophysiological testing or upon the unexpected occurrence of a very serious, and often fatal, arrhythmia.

Prediction of arrhythmogenicity.

Arrhythmogenicity sometimes is predictable by observation of drug effects on the electrocardiogram. Heart rate changes, QRS prolongation and QT interval prolongation have been suggested as warning signs but, for the most part, these reflect pharmacological effects of the administered drug and indicate plasma level related phenomena rather than individual drug "hypersensitivity".

Is arrhythmogenicity an inherent part of antiarrhythmic activity?

Many antiarrhythmic drugs interfere with conduction velocity and refractoriness, effects which will be exaggerated or modified by the underlying pathophysiological state responsible for the arrrhythmia. It is perhaps surprising that facilitation of re-entry does not occur more often than appears the case. Aggravation of automatic and triggered automatic arrhythmias is particularly a consequence of sympathomimetic drugs and digitalis. All currently known antiarrhythmic drugs can in theory and do in practice, aggravate arrhythmias.

DRUG EFFICACY VERSUS ARRHYTHMOGENICITY

Clinical studies.

Separation of lack of efficacy and arrhythmogenicity is difficult. In large population studies, detailed arrhythmia monitoring, which might give direct information on drug induced problems, is often impractical. Arrhythmogenicity may be inferred from studies in which placebo therapy is shown to be better than active therapy. However, if the drug performs better than placebo, it is rare to consider that the positive effect may have been offset in part by a negative effect reflecting drug toxicity. The possibility that some patients may be disadvantaged by drug therapy is important as it behoves us to identify the characteristics of these individuals in an attempt to protect them.

Individual patient management.

Identifying drug arrhythmogenicity when managing individual patients is similarly difficult. In patients with potentially life-threatening arrhythmias, sudden cardiac death may be considered a failure of antiarrhythmic therapy rather than as something which has been directly caused by the use of that therapy. Incrimination of the drug as responsible requires detailed knowledge of the natural history of the clinical situation, information that all too often is lacking.

CONCLUSIONS

Sudden cardiac death may be caused by a variety of ventricular tachyarrhythmias. These arrhythmias can be produced by antiarrhythmic therapy and increased awareness of the possibility and its prospective scientific investigation are mandatory if we are to profile our use and application of these agents for maximum patient benefit.

RECOMMENDATIONS.

a) Consider that arrhythmogenicity is a feature to be expected of all antiarrhythmic compounds.

b) Arrhythmogenicity should be adequately defined and sought in all antiarrhythmic drug studies.

c) Characteristics of patients dying during antiarrhythmic drug trials should be carefully analysed and arrhythmogenicity considered the cause of death until proven otherwise.

d) As invasive electrophysiological testing is the best tool for detecting arrhythmogenesis, ALL new antiarrhythmic therapies should be rigourously tested in an invasive electrophysiological protocol involving patients with life-threatening ventricular tachyarrhythmias.

e) Reports of arrhythmogenicity for ALL antiarrhythmic drugs (approved and investigational) should be collated - a politically and professionally difficult task. Is there a role for a journal devoted to brief adverse drug reaction reports?

REFERENCES.

1. Schaffer WA, Cobb LA. Recurrent ventricular fibrillation and modes of death in survivors of out-of-hospital ventricular fibrillation. New Eng J Med 293:259-262,1975
2. Liberthson RR, Nagel EL, Hirschman JC, Nussenfeld SR, Blackbourne BD, Davis JH. Pathophysiologic observations in prehospital ventricular fibrillation and sudden cardiac death. Circulation 49:790-798,1974
3. Julian DG, Valentine PA, Miller GG. Disturbances of rate rhythm and conduction in acute myocardial infarction. Am J Med 37:915-927,1964
4. Roelandt J, Klootwijk J, Lumsen J, Janse MJ. Sudden death during long term ambulatory monitoring. Eur Heart J 5:7-20,1984

5. Adgey AAJ, Allen JD, Geddes JS, James RGG, Webb SW, Zaida SA, Pantridge JF. Acute phase of myocardial infarction. Lancet 2:501-504,1971
6. Lie KI, Liem KL, Schuilenberg RM, David GK, Durrer D. Early identification of patients developing late in-hospital ventricular fibrillation after discharge from the coronary care unit. A 5.5 year retrospective study of 1897 patients. Am J Cardiol 41:674-677,1978
7. Hauer RNW, Lie KI, Liem KL, Durrer D. Long-term prognosis in patients with bundle branch block complicating acute anteroseptal infarction. Am J Cardiol 49:1581-1585,1982.
8. Geddes JS, Adgey AAJ, Pantridge JF. Prognosis after recovery from ventricular fibrillation complicating ischaemic heart disease. Lancet 2:273-275,1969.
9. McKenna WJ, England D, Doi YL, Deanfield JE, Oakley CM, Goodwin JF. Arrhythmia in hypertrophic cardiomyopathy: I Influence on prognosis. Br Heart J 46:168-172,1981.
10. Horowitz LN, Josephson ME, Kastor JA. Intracardiac electrophysiologic studies as a method for the optimization of drug therapy in chronic ventricular arrhythmia. Prog Cardiovasc Dis 23:81-98,1980.
11. Campbell RWF. Does non beta-blocking antiarrhythmic therapy prevent sudden cardiac death? In:Morganroth J, Moore EN (eds) Sudden death and congestive heart failure: diagnosis and treatment. Martinus Nijhoff, The Hague, 1983, pp140-154.
12. Myerburg RJ, Conde C, Sheps DS, Appel RA, Kiem I, Sung RJ, Castellanos A. Antiarrhythmic drug therapy in survivors of pre-hospital cardiac arrest: comparison of effects on chronic ventricular arrhythmias and recurrent cardiac arrest. Circulation 59:855-863,1979.
13. Ruskin JN, DiMarco JP, Garan H. Out-of-hospital cardiac arrest. N Engl J Med 30:607-613,1980.
14. Horn HR, Hadidian Z, Johnson JL, Vassallo HG, Williams JH, Young MD. III. Chronic tocainide therapy studies. Safety evaluation of tocainide in the American emergency use program. Am Heart J 100:1037-1040,1980.
15. Ryden L, Ariniego R, Arnman K, Herlitz J, Hjalmarson A,

Holberg S, Reyes C, Smedgard P, Swedberg K, Vedin A, Waagstein F, Waldenstrom A, Wilhelmsson C, Wedel H, Yamamoto M. A double-blind trial of metoprolol in acute myocardial infarction. Effects on ventricular tachyarrhythmias. New Engl J Med 308:614-618,1983.

16. The Norwegian Muticenter Study Group. Timolol-induced reduction in mortality and reinfarction in patients surviving acute myocardial infarction. New Engl J med 304:801-807,1981.
17. B-Blocker Heart Attack Trial Research Group. A randomized trial of propranalol in patients with acute myocardial infarction. JAMA 247:1707-1714,1982.
18. Hjalmarson A, Herlitz J, Malek I, Ryden L, Vedin A, Waldenstrom A, Wedel H, Elmfeldt D, Holmberg S, Nyberg G, Swedberg K, Waagstein F, Waldenstrom J, Wilhelmsen L, Wilhelmsson C. Effect on mortality of metoprolol in acute myocardial infarction. A double blind randomized trial. Lancet 2:823-827,1981.
19. Schwartz PJ. The idiopathic long QT syndrome. The need for a prospective registry. Eur Heart J 4:529-531,1983
20. McKenna WJ. Arrhythmia and prognosis in hypertrophic cardiomyopathy. Eur Heart J 4(suppl F):225-234,1983.
21. Winkle RA, Mason JW, Griffin JC, Ross D. Malignant ventricular tachyarrhythmias associated with the use of encainide. Am Heart J 102:854-864,1981.
22. Nicholson MR, Campbell RWF, Julian DG. Ventricular tachycardia - management with encainide.(abs) Circulation 64:37,1981
23. Kennelly BM. Comparison of lidoflazine and quinidine in prophylactic treatment of arrhythmias. Br Heart J 39:540-546,1977.
24. Ruskin JN, McGovern B, Garan H, DiMarco JP, Kelly E. Antiarrhythmic drugs: a possible cause of out-of-hospital cardiac arrest. N Engl J Med 309:1302-1306,1983.
25. Denes P, Gabster A, Huang SK. Clinical, electrocardiographic and follow-up observations in patients having ventricular fibrillation during Holter monitoring. Role of quinidine

therapy. Am J Cardiol 48:9-16,1981.

26. McGovern B, Garan H, Kelly, Ruskin JN. Adverse reactions during treatment with amiodarone hydrochloride. Br Med J 287:175-180,1983

27. Nathan AW, Hellestrand KJ, Bexton RS, Banim SO, Spurrell RAJ, Camm AJ. Proarrhythmic effect of the new antiarrhythmic agent flecainide acetate. Am Heart J 107:222-228,1984.

13

SUDDEN DEATH AS AN END-POINT FOR THE CLINICAL EVALUATION OF ANTIARRHYTHMIC DRUGS

Jay N. Cohn, M.D. and Thomas Rector, Ph.D.

The goal of antiarrhythmic therapy in patients with ventricular premature beats or runs of ventricular tachycardia is usually aimed at preventing dysrhythmia induced death rather than alleviating symptomatology. The major dilemma facing the clinician is that the use of easily quantifiable total or partial pharmacologic suppression of ventricular arrhythmia during a period of electrocardiographic monitoring has not been shown to be associated with a subsequent reduction in the risk of dysrhythmic deaths. Thus, it is appropriate to consider whether sudden death can be adequately identified as a dysrhythmic event so that it can be used as an end-point for a clinical trial evaluating antiarrhthymic agents.

Sudden Death as an Antiarrhythmic End-point

In any mortality study total cardiac deaths must serve as an end-point. Analysis of cardiac death merely requires censoring of those deaths clearly due to non-cardiac causes. Although the distinction sometimes becomes difficult in individual cases, the separation into cardiac and non-cardiac deaths should be quite accurate. Subdivision of cardiac deaths into dysrhythmic and non-dysrhythmic causes is considerably less rigorous. Nontheless, the need to establish the antiarrhythmic efficacy of new drugs may make it necessary to undertake such a classification. For example, it is possible that an antiarrhythmic agent with negative inotropic properties might reduce the risk of dying from a dysrhythmia but increase the risk of dying from heart failure. The absence of an effect on total cardiac mortality might imply that no effect had been noted, whereas the redistribution of cause of death might have great importance by encouraging the concept of multiple pharmacologic appraoches, each to counteract another mechanism that may be involved in the occurrence of cardiac death.

As an end-point to evaluate antiarrhythmic efficacy 'sudden' cardiac death ideally would be defined in a manner that reliably identifies deaths primarily or secondarily due to a dysrhythmia that if successfully treated would prolong life. Can these dysrhythmic deaths be reliably and validly identified among all the deaths that occur? Most individuals who experience life-threatening arrhythmias also have ischemic heart disease or some form of cardiomyopathy (1). The use of a time criterion, e.g. death within one hour of the onset of symptoms, does not exclude those who expire from massive myocardial damage or other acute cardiovascular events. A strict time criterion may also be inappropriate for a patient who sustains a cardiac arrest but is resuscitated to survive several days in a comatose state before expiring.

Unwitnessed deaths present a problem that can only be incompletely resolved. Enough history usually can be obtained to make a reasonable judgement as to the apparent health of the individual when last seen, the time interval between the observation and his death, and his activity at the time of death. Although sudden unexpected death could be caused by many catastophic episodes other than a ventricular arrhythmia, the relative likelihood of this being at least a partial cause in the type of high-risk population to be entered in such a study would probably render the error rate in labeling such deaths as "presumed arrhythmic" given no evidence to the contrary low enough to be acceptable.

Much emphasis has been placed on the presence of premonitory symptoms or of pathological evidence of fresh myocardial infarction as criteria for the classification of sudden death (2). However, the important issue is not whether the arrhythmia was precipitated by an ischemic episode but whether prevention of the arrhythmia could have prolonged life. Conclusions regarding this latter query become particularly difficult when dealing with individuals who have severe progressive heart failure but eventually expire in the hospital from a sudden arrhythmia. With these patients, of course, prevention of the terminal arrhythmia may not significantly prolong life. These deaths might better be classified as "pump failure" rather than "dysrhythmia".

For the purpose of descritive classification we have found it useful to divide cardiac deaths into four groups: (1) Sudden and unexpected death presumably due to a dysrhythmia. Temporary resuscitation does not exclude the patient from this category. In this group prevention of the arrhythmia should have the greatest life-prolonging effect; (2) presumed arrhythmic deaths occuring in the setting of signs or symptoms of unstable heart disease. The assumption in this category is that the presumed dysrhythmia probably shortened the patient's life; (3) expected deaths due to severe and unstable heart disease in whom it is not thought that dysrhythmia control could have substantially prolonged life; (4) other deaths probably or possibly related to the heart disease, e.g. stroke. It is of vital importance that the classification of individual cases be carried out with as much systematic data on the terminal illness as can be obtained but with the classifiers totally blinded to any treatment information.

Patient Selection

In order to carry out an antiarrhythmic trial it is necessary to select a patient population in whom the arrhythmia related event rate is high enough to make the study feasible. Risk factors that may be used to select or stratify a study sample are left ventricular functional impairment, the presence of complex ventricular arrhythmias demonstrated by continous monitoring, the presence of ischemia in patients with coronary disease on the basis of stress testing, and proximity to a prior acute myocardial infarction or resuscitated cardiac arrest (3-7).

Although most studies have utilized patients with coronary disease the possibility of also using a patient population with primary cardiomyopathy should be considered. Arrhythmias are thought to be a common terminal event in patients with congestive cardiomyopathy (8). Since these individuals are not at risk for ischemic episodes with their attendent electrical instability, the patient with primary myocardial disease might serve as a purer subpopulation to test the antiarrhythmic efficacy of various drugs. It must be recognized, however, that the cardiac mortality rate in this patient population may be slightly lower then the group with ischemic heart disease and thus a larger sample size may be necessary for such a trial (9). In addition, one could not casually extrapolate such findings to individuals with ischemic disease.

Drug selection

Although as previously stated the aim of antiarrhythmic therapy is to prevent death, most previous trials have utilized a reduction in premature ventricular depolarizations as the therapeutic goal of the study. Therefore, drugs employed for this purpose in the past have of necessity been agents which suppress ventricular premature depolarizations. Using cardiac death or presumed arrhythmic death as end-points for a clinical trial opens the possibility for using a variety of drugs that do not necessarily suppress ventricular premature beats but may nonetheless have a favorable effect on the incidence of dysrhythmic death. At least four classes of agents might be considered:

1. Antiarrhythmic agents: These might include various type 1 antiarrhythmic drugs which are already undergoing

considerable clinical trial or some of the newer type 3 agents which might have a particularly beneficial effect on dysrhythmic death (10).

2. Antifibrillatory agents: Drugs such as bethanidine and bretylium have been classified by some investigators as antifibrillatory because of their effect on raising the fibrillation threshold in experimental animals even though they do not necessarily inhibit premature ventricular depolarizations (11).

3. Anti-ischemic agents: If episodes of myocardial ischemia are critical for the development of electrical instability and cardiac death an agent which could prevent the development or severity of ischemia might have a favorable effect on the incidence of arrhythmic death. The success of beta-blockers in reducing mortality in the first year after acute myocardial infarction could represent an example of an anti-ischemic drug having an effect on the incidence of cardiac death (12). This mechanism for the efficacy of beta-blockers in this clinical situation has not been established. If an anti-ischemic drug, such as a beta-blocker, a calcium antagonist or a nitrate, would be used in a clinical trial then the patient population might appropriately be restricted to patients with coronary artery disease. Also nondysrhythmic and dysrhythmic categorizations of cardiac death should be examined at least descriptively.

4. Drugs to improve left ventricular function: Since the severity of left ventricular dysfunction is closely

correlated with the risk of cardiac death a drug designed to improve the performance of the left ventricle could be an effective way to prevent dysrhythmic death (13). The use of drugs that reduce aortic impedance, reduce preload or increase contractility all could serve as appropriate agents to test the efficacy of this approach to prevention of cardiac death. Use of a population with cardiomyopathy in the absence of coronary disease would provide a purer substrate in which to test this hypothesis since influences on ischemia need not be considered in this population.

Sample size

In designing an intervention trial in a patient population at high risk to experience dysrhythmic death, consideration must be given to the effect of end-point selection on sample size. Estimates must be based on knowledge about the event rate in the population (cardiac death or dysrhythmic death) and the presumed efectiveness of the therapy. If a therapy were effective only on dysrhythmic deaths a smaller sample might be necessary to demonstrate a prolongation of time to dysrhythmic deaths than to total cardiac deaths, but the smaller event rate and need to censor all deaths not due to dysrhythmias might counterbalance this advantage. Any inaccuracy in the classification of dysrhythmic death might further reduce the power of the study.

How many subjects would be required for an antiarrhythmic trial when using all cardiac deaths versus only dysrhythmic cardiac deaths as an end-point? Assume a study sample which experiences 30% annual cardiac mortality is accrued over a year and followed for an additional year. Also assume that half the cardiac deaths would be classified as dysrhythmic end-points, the survival curves are exponential, treatments are allocated equally, a two-sided significance test is used with alpha equal to 0.05 and a power of 90%, and the antiarrhythmic agent would reduce so-called dysrhythmic deaths by 50% during the first year compared to placebo without preventing or causing any of the other cardiac deaths. The estimated total sample size using cardiac death as an end-point would be approximately 1050 compared to about 525 using the dysrhythmic death end-point. Neither estimate incorporates censoring due to withdrawals, losses to follow-up, or deaths other than defined by the end-point of interest. The sample size could be reduced by prolonging the follow-up period. Looking at these estimates in other terms, the expected median time for total cardiac death would be prolonged by 40% whereas the median time to dysrhythmic death would be prolonged by more than 100%.

Conclusion

Mortality should be the end-point for studying antiarrhythmic agents. By careful patient selection and uniform criteria for classifying dysrhythmic deaths, it should be possible to design trials of moderate size that could establish the efficacy of a variety of pharmacologic agents in this high-risk patient population.

REFERENCES

1. Graboys TB, Lown B, Podrid RJ, DeSilva R: Long-term survival

of patients with malignant ventricular arrhythmia treated with antiarrhythmic drugs. Am J Cardiol 50: 437-443, 1982.

2. Hinkle LE, Thaler HT: Clinical classification of cardiac deaths. Circulation 65: 457-464, 1982.
3. Epstein SE, Palmere ST, Patterson RE. Evaluation of patients after acute myocardial infarction. N Engl J Med 50: 437-443, 1982.
4. Weaver WD, Cobb LA, Hallstrom AP: Ambulatory arrhythmias in resuscitated victims of cardiac arrest. Circulation 66: 212-218, 1982.
5. Goldstein S, Landis R, Leighton R, Ritter G, Vasu M, Lantis A, Serokman R: Characteristics of the resuscitated out-of-hospital cardiac arrest victim with coronary heart disease. Circulation 64: 977-984, 1981.
6. Ruberman W, Weinblatt E, Goldberg JD, Frank CW, Shapiro S: Ventricular premature beats and mortality after myocardial infarction. New Engl J Med 297: 750-757, 1984.
7. Schaffer WA, Cobb LA: Recurrent ventricular fibrillation and modes of death in survivors of out-of-hospital ventricular fibrillation. N Engl J Med 293: 259-262, 1975.
8. Cohn JN, Levine TB, Oliveri MT, Garberg V, Lura D, Francis GS, Simon AD, Rector T: Plasma norepinephrine as a guide to prognosis in patients with chronic congestive heart failure. N Engl J Med 311: 819-823, 1984.

9. Franciosa JA, Wilen M, Ziesche S, Cohn JN: Survival in men with severe chronic left ventricular failure due to either coronary heart disease or idiopathic diluted cardiomyopathy. Am J Card 51: 831-836, 1983.

10. Hamer AWF, Peter T, Weiss D, Mandel WJ: The empirical use of chronic oral amiodarone therapy to prevent recurrent, sudden cardiac death. New Aspects in the Medical Treatment of Tachyarrhythmias, Breithardt G and Loogen F (eds.), Urban and Schwarzenberg, 1983, 153-157.

11. Bacaner MB, Benditt DG: Antiarrhythmic, antifibrillatory, and hemodynamic actions of bethanidine sulfate: an orally effective analog of bretylium for suppression of ventricular tachyarrhythmias. Am J Cardiol 50: 728-734, 1982.

12. Beta-blocker Heart Attack Trial Research Group: A randomized trial of propranolol in patients with acute myocardial infarction. I. Mortality Results. JAMA 247: 1707-1714, 1982.

13. Swerdlow CD, Winkle RA, Mason JW: Determinants of survival in patients with ventricular tachyarrhythmias. N Eng J Med 308: 1436-1442.

14

HOLTER/EXERCISE AND ELECTROPHYSIOLOGIC METHODS FOR EVALUATING DRUG THERAPY FOR MALIGNANT VENTRICULAR ARRHYTHMIAS: DO WE NEED BOTH MODELS?

J. THOMAS BIGGER, JR., M.D.

Many studies have shown the malignant potential of symptomatic sustained ventricular arrhythmias, i.e., ventricular tachycardia (VT) or ventricular fibrillation (VF). Routine clinical evaluation usually reveals ample evidence for the existence of heart disease in patients with these arrhythmias. Coronary heart disease is by far the most frequently occurring etiologic form of heart disease (about 80%) followed by cardiomyopathy and valvular heart disease. Often ventricular scarring and/or hypertrophy is present in patients with sustained VA. Left ventricular dysfunction is apparent clinically as heart failure or detectable by radionuclide ventriculography. A 24-hour ECG recording frequently will reveal repetitive ventricular arrhythmias, including runs of unsustained VT. When antiarrhythmic drug treatment is stopped, electrophysiologic studies are capable of inducing sustained VT or VF in about 80% of those who present with recurrent sustained VT as the clinical arrhythmia and in about 70% of those who had a cardiac arrest and were found in VF (1-4).

There is a strong consensus that malignant ventricular arrhythmias should be treated vigorously and that treatment should be evaluated using precise endpoints. There are two methodologies that are reasonably accurate for predicting the long-term outcome of antiarrhythmic treatment: (1) a noninvasive approach using Holter ECG recordings and exercise testing (5), and (2) an electrophysiologic approach using programmed ventricular stimulation (1-4). The Holter/exercise method and the electrophysiologic method both are complex and time consuming. Both require a high degree of expertise. There has been no direct comparison of the two methods so that a preference cannot be stated in most cases. However, there are certain circumstances in which one or the other method is clearly preferred to the other. For cases where VT is incessant or sustained VT occurs frequently but does not cause severe hemodynamic effects, the Holter/exercise approach is preferable. Such cases represent less than 5% of those encountered on most busy arrhythmia units. In cases where the frequency of occurrence of sustained VT is low and its effects are devastating, the electrophysiologic approach is considered superior. This type of case represents about 20% to 25% of those managed on an arrhythmia service.

Holter/exercise method. The Holter/exercise method as proposed by the arrhythmia service at the Brigham and Women's Hospital is outlined in Table 1. This group reported the long-term outcome of patients with malignant ventricular arrhythmias who were managed using the Holter/exercise approach in 1982 (5). Two other groups have used Holter recordings in a similar way but have not used exercise to assess antiarrhythmic efficacy (6,7). In the study by Graboys et al. and by Vlay et al., efficacy was defined as elimination of unsustained VT from 24-hour Holter ECG recordings (6,7). Hoffman et al. defined efficacy as elimination of unsustained VT and paired VPD from 24-hour Holter ECG recordings (7). Table 2 shows that the probability of dying during one year of follow-up was much greater in the patients whose unsustained VT could not be completely suppressed. The odds of dying for patients who did not respond to treatment were 6 to 17 times as great as for patients whose unsustained VT was totally suppressed by treatment. Because there is no controlled comparison to an untreated group we cannot attribute the improved survivorship in patients who respond to drug therapy to a drug effect on mortality. It is possible that the response to drug therapy separates the good from poor risk patients without having any effect on outcome. However, these data do suggest that the Holter/exercise method of Graboys et al. or just Holter evaluation is useful for predicting the outcome of antiarrhythmic treatment in patients with malignant ventricular arrhythmias. Also, these data suggest that 60 to 80% of patients treated can be controlled as judged by this method.

Electrophysiologic method. The electrophysiologic method is outlined in Table 3. Several groups have reported on the ability of programmed ventricular stimulation to guide drug therapy. As an example,

Table 1. Holter/Exercise Approach for Management of Malignant Ventricular Arrhythmias.

PHASE 0: Baseline (1)

- Admit to CCU
- Withdraw Antiarrhythmic Drugs
- 2 24-Hour ECG Recordings
- Treadmill Exercise (Bruce Protocol)

PHASE 1: Acute Drug Tests (6)

- Single Large Oral Dose of Drug
- 6 Hours of Trendscription
- 6 Hourly Exercise Tests
- 6 Hourly Serum Drug Tests
- Acetylstrophanthidin Test

PHASE 2: Steady-State Drug Tests (4)

- Repeated Drug Doses (48 Hours)
- 24-Hour ECG Recording
- Exercise Tests
- Serum Drug Tests

PHASE 3: Follow-up (1)

Treatment

- Antiarrhythmic Drug
- Beta Adrenergic Blocking Drug
- Digoxin

Evaluation

- 24-Hour ECG Recording
- Exercise Test
- Serum Drug Test

Numbers in parentheses indicate the average number of times that a step is repeated.

data from Swerdlow et al. at Stanford University is shown in Table 4 (8). The endpoint for judging drug effectiveness was a change from sustained VT or VF to less than five repetitive ventricular responses in response to programmed ventricular stimulation. The odds of dying for drug failures is eight times greater than for drug successes. The 95% confidence limits for the odds ratio are 4.3 to 15.0; therefore, this result is not significantly different from that obtained with the

Table 2. Holter/Exercise Approach to Malignant Ventricular Arrhythmias.

Author	Patients		Success		Mortality Rate	
	No.	Type	No.	%	Success	Failure
Graboys et al.[5]	123	VT-H, VT-Sy, VF	98	80%	5%	50%
Vlay et al.[6]	44	VT-S, VF	26	59%	0%	33%
Hoffman et al.[7]	50	VT-U	39	78%	7%	36%

VT-H = Hemodynamically compromising VT; VT-Sy = VT with syncope;
VT-S = sustained VT; VT-U = unsustained VT; VF = ventricular fibrillation.

Table 3. Electrophysiologic Approach to Malignant Ventricular Arrhythmias.

1. Stabilize Patient; Treat Complications of Arrest
2. Continuous ECG Monitoring in Intensive Care or Telemetry Unit
3. Withdraw Antiarrhythmic Drugs
4. Determine Inducibility using Programmed Ventricular Stimulation
5. Therapy
   - -- Antiarrhythmic Drug
   - -- Arrhythmia Surgery
   - -- Anti-Tachycardia Pacemaker
   - -- Automatic Implanted Defibrillator
   - -- Stop Antiarrhythmic Drug
6. Determine Lack of Inducibility using Programmed Ventricular Stimulation

Holter/exercise approach. Since these electrophysiologic studies are not controlled either, the same remarks made above about the Holter/exercise method also pertain to the electrophysiologic approach. That is, it is possible that drug responsiveness can predict outcome without directly influencing it. The proportion of patients with sustained ventricular arrhythmias that can be controlled is about 50% when

Table 4. Electrophysiologic Studies (EPS) for Malignant Ventricular Arrhythmias.

EPS Prediction	No.	Three Year Mortality Rate
Success	103	20%
Faiure	102	68%

judged by the electrophysiologic method.

It is apparent from the results given above that either the Holter ECG-exercise approach or clinical electrophysiologic studies can serve as a guide to therapy and that either is preferable in specific subgroups of patients. Are these two methods more or less equivalent? Or, is one better than the other in patients that are subject to management with either method?

Are the two methods equivalent?

The electrophysiologic method identifies a successful drug response to a class 1 agent about 30% of the time. Electrophysiologic methods show that about 20% of the patients who fail all class 1 drugs will respond to amiodarone. Initial studies concluded that the efficacy of amiodarone could not be accurately evaluated using electrophysiologic methods (9-11). More recently, it has been found that electrophysiologic studies do predict the outcome on amiodarone with some accuracy if they are performed long enough after exposure to the drug (12-14). The non-invasive approach identifies an effective drug regimen in almost 80% of a population of patients with malignant ventricular arrhythmias. The marginal proportions in Table 5 summarize the preliminary estimates of concordance between the two methods obtained by small studies in different populations. The proportions in the cells of Table 5 are hypotheses about the relationships between the Holter/exercise and electrophysiologic methods in a single population. The latter estimates assume that previous studies have been done in equivalent populations. Looking at these figures, it appears that patients with malignant

Table 5. Different Success Rates for Holter/Exercise and Electrophysiologic Studies (EPS).

	Holter/Exercise		
EPS	Success	Failure	Total
Success	0.40	0.10	0.50
Failure	0.40	0.10	0.50
Total	0.80	0.20	1.00

ventricular arrhythmias are significantly more likely to be judged efficacious by noninvasive methods than by electrophysiologic studies. The first question may be: is this difference real? It is possible that the patient populations reported by groups who use noninvasive methods to evaluate antiarrhythmic treatment are significantly different from those reported by groups who use electrophysiologic methods. For example, community physicians may be more likely to refer sicker patients or those whose arrhythmia is judged to be more severe for electrophysiologic studies. This practice would be based on the concept that electrophysiologic studies are more definitive and should be used in preference to noninvasive methods in a sicker group. Or, conversely, that patients who are not at such high risk can be managed by noninvasive methods. It is likely that such biases in referral patterns do exist. A review of the proportion of patients with coronary heart disease, the degree of left ventricular dysfunction and the severity and chronicity of the presenting arrhythmia in the report of Graboys et al. suggests that their patients are quite similar to those reported by groups that have used electrophysiologic studies to evaluate antiarrhythmic drug treatment of malignant ventricular arrhythmias. Until patients are characterized using a standardized set of criteria and scales, we will not be able to compare severity of disease among different studies. Some of the criteria that might be used to characterize groups of patients in order to judge their comparability are given in Table 6.

Table 6. Variables that Should Be Compared among Studies of Malignant Arrhythmias.

ETIOLOGIC HEART DISEASE

-- Coronary Heart Disease
-- Cardiomyopathy
-- Other

LEFT VENTRICULAR DYSFUNCTION

-- NYHA Functional Class
-- Duration of Heart Failure
-- Presence of LV Aneurysm
-- LV EDP
-- LV EDV
-- LV Ejection Fraction
-- Exercise Performance
-- Plasma Norepinephrine

ARRHYTHMIA HISTORY

-- Type: VT-U; VT-S; VF
-- Symptomatic: Mild; Moderate; Severe
-- Number of Episodes
-- Time since First Episode
-- Number of Unsuccessful Drug Trials

What accounts for the difference between Holter/exercise and electrophysiologic studies in predicting treatment outcome?

A second question is: if a real difference in success rate exists between noninvasive methods and electrophysiologic studies, what accounts for it? In 95% of the cases managed by noninvasive methods, sustained VT is not the arrhythmia that is used to judge the efficacy of treatment. Much more often VPD frequency and repetitive activity, i.e., runs of unsustained VT are used to gauge antiarrhythmic treatment. One may view the patient with a large ventricular scar and intermittent sustained VT as having an anatomically determined reentrant circuit that can support sustained ventricular arrhythmias under appropriate conditions. Sporadic VPD may be viewed as triggering events that are capable of provoking sustained reentrant activity in the circuit. Repetitive VPD can be thought of in several different ways with respect to this schema. Repetitive VPD could be considered extreme forms of the triggering mechanism. We know that a single premature response invoked by electrical stimulation rarely can provoke sustained VT in susceptible patients even when the diastolic interval is carefully scanned. This will occur in 10% or fewer of the patients who are inducible with more aggressive protocols (15). However, double or triple premature responses are 70% (double) or 90% (triple) likely to induce sustained ventricular arrhythmias. Thus, short runs of unsustained VT may have a fundamentally different mechanism from that of the sustained VT and act to provoke sustained VT. Another way of thinking about spontaneous unsustained VT is that the repetitive activity uses the reentrant circuit in scarred ventricles but indicate that conditions are not quite right to support spontaneous sustained VT at that time. During electrophysiologic studies, the heart rate usually is changed by atrial or ventricular pacing or by isoproterenol infusion and a wide variety of premature stimulation patterns are used to provoke sustained VT. It may be that the brief runs of spontaneous repetitive responses indicate that the appropriate conditions of refractoriness and conduction delay required to achieve sustained VT have not been produced by the spontaneous ventricular activity. Spontaneous VPD rarely occur as early as the stimuli used in electrophysiology laboratory protocols. Also, if spontaneous, repetitive VPD use the anatomic reentrant circuit, these spontaneous repetitive events interact with the anatomic circuit in a way quite different from repetitive VPD invoked by electrical stimulation even when the interectopic intervals are similar. In the latter case, two or three premature responses are evoked outside the circuit and have to propagate into it. Thus, it is quite possible that drugs could alter both the tissues in the circuit and those outside in such a way that the probability of spontaneous sustained VT would be extremely low while it was still possible to provoke sustained VT with programmed ventricular stimulation.

A comparison is needed!

Because of the potential differences in patient characteristics among previous studies of sustained ventricular arrhythmias, a direct comparison will be necessary to determine whether there is a systematic difference between the Holter/exercise method and electrophysiologic approach.

Table 7. Objectives of a Trial to Compare the Holter/Exercise Method (H-Ex) to Electrophysiologic Studies (EPS).

1. Determine the proportion of patients who have malignant ventricular arrhythmias who are subject to study by H-Ex and by EPS.

2. Compare the long-term accuracy of efficacy prediction by H-Ex and EPS.

   a. Total mortality
   b. Sudden cardiac death
   c. Arrhythmic death (Hinkle)
   d. Recurrence of symptomatic arrhythmias

3. Compare the probability of predicting efficacy with H-Ex and EPS

4. Compare the duration of hospital stay, costs, etc.

5. Determine the extent of agreement between H-Ex and EPS

6. Determine drug specific efficacy by H-Ex and EPS

Some of the objectives of the trial are shown in Table 7. The primary purpose of the trial would be to compare the accuracy of the Holter/exercise method and electrophysiologic technique to predict long-term effective antiarrhythmic treatment. To make the comparison valid and convincing, clear-cut end points must be used. It would be desirable if arrhythmic death and symptomatic recurrences of sustained ventricular arrhythmias could be combined as the primary end point. However, the assignment of deaths to pathophysiologic subgroups is subject to much uncertainity and to bias. Therefore, the primary mortality end point should be total mortality. The primary arrhythmic end point should be documented recurrences of sustained ventricular arrhythmias. Secondary end points can be sudden cardiac death, arrhythmic death by the Hinkle classification and the quantitation of repetitive VPD and VPD frequency in 24-hour ECG recordings. A second important objective of a randomized study using the Holter/exercise and electrophysiologic approach is to provide insight into other significant issues, e.g., comparative costs, personnel skills and resources and drug efficacy. Several technical problems plague both methods. Electrophysiologic methods are not standardized and preferences among alternate methods have not been clarified. Similarly, the accuracy of commercial Holter systems is yet to be determined. Such studies as have been done suggest that significant errors may exist in the detection and quantitation of repetitive ventricular arrhythmias in 24-hour ECG recordings (16,17). Results with research Holter analysis systems are more encouraging (18). A comparative trial would provide some valuable information on these technological issues. Depending on the design, some useful information on the comparative efficacy of antiarrhythmic drugs against sustained ventricular arrhythmias might emerge also.

For several reasons, I think that a trial to compare the Holter/exercise approach and electrophysiologic studies should be a multicenter effort. The multicenter approach should protect against several sources of bias, e.g., type of hospital, region of the country, local referral patterns and local practice customs. Also, a multicenter study can provide an adequate sample size in a reasonable period of time. Finally, a multicenter study is likely to have more scientific and clinical credibility and acceptability.

REFERENCES

1. Josephson ME, Horowitz LN: Electrophysiologic approach to therapy of recurrent sustained ventricular tachycardia. Am J Cardiol. 43:631-642, 1979.

2. Mason JW, Winkle RA: Accuracy of the ventricular tachycardia-induction study for predicting long-term efficacy and inefficacy of antiarrhythmic drugs. N Engl J Med. 303:1073-1077, 1980.

3. Horowitz LN, Josephson ME, Kastor JA: Intracardiac electrophysiologic studies as a method for the optimization of drug therapy in chronic ventricular arrhythmia. Prog Cardiovasc Dis 23:81-98, 1980.

4. Ruskin JN, DiMarco JP, Garan H: Out-of-hospital cardiac arrest. Electrophysiologic observations and selection of long-term antiarrhythmic therapy. N Engl J Med 303:607-613, 1980.

5. Graboys TB, Lown B, Podrid PJ, DeSilva R: Long-term survival of patients with ventricular arrhythmia treated with antiarrhythmic drugs. Am J Cardiol 50:437-443, 1982.

6. Vlay SC, Kallman LH, Reid PR: Prognostic assessment of survivors of ventricular tachycaridia and ventricular fibrillation with ambulatory monitoring. Am J Cardiol 54:87-90, 1984.

7. Hoffman A, Schutz E, White R, Follath F, Burckhardt D: Suppression of high-grade ventricular ectopic activity by antiarrhythmic drug treatment as a marker for survival in patients with chronic coronary artery disease. Am Heart J 107:1103-1108, 1984.

8. Swerdlow CD, Winkle RA, Mason JW: Determinants of survival in patients with ventricular tachyarrhythmias. N Engl J Med 308:1436-42, 1983.

9. Heger JJ, Prystowsky EN, Zipes DP: Clinical efficacy of amiodarone in treatment of recurrent ventricular tachycardia and ventricular fibrillation. Am Heart J 106: 887-894, 1983.

10. Waxman HL: The efficacy of amiodarone for ventricular arrhythmias cannot be predicted with clinical electrophysiological studies. Internat J Cardiol 3: 76-80, 1983.

11. Peter T, Hamer A, Mandel WJ, Weiss D: Evaluation of amiodarone therapy in the treatment of drug-resistant cardiac arrhythmias: long-term follow-up. Am Heart J 106: 943-950, 1983.

12. McGovern B, Ruskin JN: The efficacy of amiodarone for ventricular arrhythmias can be predicted with clinical electrophysiological studies. Intern J Cardiol 3: 71-76, 1983.

13. Borggrefe M, Breithardt G, Seipel L: Value of serial electrophysiological testing in the treatment of ventricular tachyarrhythmias with amiodarone (Abstr). Circulation 68: III-381, 1983.

14. Horowitz LN, Greenspan AM, Spielman SR, Webb CR, Morganroth J, Rotmensch H, Sokoloff NM, Rae AP, Segal BL, Kay HR: Usefulness of electrophysiologic testing in evaluation of amiodarone therapy for sustained ventricular tachyarrhythmias associated with coronary heart disease. Am J Cardiol, In press, February 1985.

15. Livelli FD, Bigger JT Jr, Reiffel JA, Gang ES, Patton JN, Noethling PM, Rolnitzky LM, Gliklich JI: Response to programmed ventricular stimulation: sensitivity, specificity and relationship to heart disease. Am J Cardiol 50: 452-458, 1982

16. Oliver GC, Ripley KL, Miller JP, Thomas LJ Jr: A critical review of computer arrhythmia detection. In: L Pordy (ed) Computer electrocardiography: current status and criteria. Futura Press, Mount Kisco, NY, 1977, pp 319-325.

17. Hermes RE, GC Oliver: Use of the American Heart Association database. In: NK Wenger, MB Mock, I Ringqvist (eds) Ambulatory electrocardiographic recording. Year Book Medical Publishers, Inc, Chicago, 1981, pp 165-181.

18. Clark KW, Rolnitzky LM, Miller JP, DeCamilla JJ, Kleiger RE, Thanavaro S, Bigger JT Jr and other MPIP participants: Ambulatory ECG analysis shared by two independent computer labs in the multicenter post-infarction program (MPIP). In: Computers in cardiology 1980, IEEE Computer Society, Long Beach, CA, 1981, pp 271-275.

PANEL ON SPECIAL CONSIDERATIONS AND FDA STANDARDS

Dr. Harrison: We talked about the individual versus the group response to antiarrhythmic drug therapy and there are some other statistical considerations and design considerations. Dr. Lipicky, how do you see this from the standpoint of FDA guidelines. Do you want both of these or would you rather have responses that are for individual patients and with those individual responses defined in a very specific way or are you looking at group responses or do you need both?

Dr. Lipicky: The obvious answer I think is both. The interesting thing is that there has been relatively limited experience I think where things such as dose response were looked at on the basis of both individual data and then in some fashion group data. The interesting thing to me has been where both of those things have been applied. That for some reason or another the group data kind of looks like the individual data, and that there really aren't enormous differences that come from that. It makes it a fairly persuasive argument when one looks at say individual data and then makes some predictions from that and confirms that in group data. You can actually pull information together that otherwise couldn't be pulled together. For example, if there were a number of different studies, all independent, individual studies that say use different doses and one had some individual data where individuals received more than one dose, you could then get a feeling for what the dose response relationship was in individuals and then when you look across studies and found that that looked pretty much the same, that is a fairly persuasive set of data, I think. It really does need to be looked at in both fashions.

Dr. Morganroth: Let me see if I can focus the question to the issue Dr. Pledger was addressing earlier and that is, if we have data on a placebo versus antiarrhythmic drug x, and then we have 100 patients on placebo and 100 patients on drug x. If you take an analysis of variance approach or some sort of group data approach concerning all the patients on placebo and compare it to the quantitation of the data on x, I think that Dr. Pledger was saying that this is the preferable statistical way. My trouble with that approach has always been, that we know as Dr. Campbell pointed out that there are going to be some patients whose PVC frequency goes up, other patients in whom there is a reduction and some will have no effect. It seems to me if you just look at the group effect and if a drug causes a lot of proarrhythmia, you might automatically assume that that drug was not different from placebo where in fact maybe a quarter of the patients did well and one quarter had no change but half got worse, using group data you would miss the point where you wouldn't be using individual data. I grant you with individual data, you are arbitrarily picking a percent efficacy or threshold of all-or-nothing response, but as I pointed out I think you should choose many levels. You can pick 80%, 90% or 100% or less reduction.

Dr. Pledger: I guess I don't completely agree. In the study you describe, I do think that the group analysis is the primary one. It is the most relevant and is meaningful. If in fact you were unfortunate or uninformed going into this study and selected patients who would respond and patients who would get a very bad outcome, then it is true that the means may not be different. The adjusted means or whatever you look at may not be different. In that case you would certainly want to know why. Essentially your study has failed I would say and you don't convert it to a successful study by saying, here are some patients who results look good, and so the drug works for those and here are some whose results look bad and the drug doesn't work for those. You say that, you look at the data and find those patients but that is not for the purpose of reaching a conclusion. To me in trying to explain what was wrong with your study design and I do recognize the desire to look at the individual patients, but I guess I am just not persuaded that you have the certainty, when you look at one patient and say this is a drug effect, that you are sure that is what it is.

Dr. Sami: I just want to qualify a point. I think in order that we agree or disagree on something, we should at least agree on definitions. If we are looking at a pharmacological action of a drug, meaning its statistically significant pharmacologic effect, then any model is O.K. so long as you can prove it statistically. For example, using group effects, you can compare a placebo to a drug trial and you can see for example that a 50% reduction in the PVC in the overall group is statistically significant, different from the null hypothesis of happening completely by chance. Now that may be totally unacceptable on a clinical level to an individual physician when he treats his patient. Maybe that 50% reduction even though it may satisfy the statistician, is not at all adequate to control let's say symptoms or prevent sudden death in the case of antiarrhythmic drugs. So I think it is important if you look at the individual results you have to define what is the variability of the test that you are using to analyze the data and compare that variability if you like, or compare that patient under placebo control and what is his variability as defined against the intervention. Now this intervention, you may find that statistically a reduction of 70% for example is statistically significant. What does it really mean to the clinician? That is an entirely different question.

Dr. Temple: These are not either/or things. As Dr. Pledger said, you can take a look at individual responses and analyze them in a grouped way. You can ask how many people had a 90% reduction, how many had an 80%. When you see a difference in distribution between your placebo group and your treatment group, you have learned something, but I think it is probably true that you would have difficulty saying for any single individual patient, this reduction was guaranteed to be due to the drug. On the other hand, I think for patient populations where the stability of their arrhythmias is reasonably well known, there is some evidence that you can say that a certain degree of change is very unlikely to be a spontaneous change. One of the nice things about antiarrhythmic drugs that many of them produce very large changes so that you have in some senses the luxery of using less rigorous study designs. We have in the past accepted things such as off-on-off designs where you plainly don't control placebo effects or investigator observational effects and all that because we think the measurements are objective and not subjective bias, because there is some automatic reading of the Holter whereas we are uncomfortable with that kind of study design in an angina study. With all that, it seems to me in some circumstances at least, by setting a very high standard for reaching a conclusion that an effect seen was a drug effect, like an 80% reduction or something like that, I'm not sure you can't rely on individual results without having in all cases a concurrent control. I mean I I think that is subject to discussion, but those kinds of changes are unusual spontaneously and are probably best attributed to drug. Just one other thought. There is one other study design that we have seen a lot of that gives you some clue as to how reproducible these things are. If you open titrate patients on the drug in question to so-called optimal dose and then randomly assign people to the optimal dose or to placebo and at least for the tocainide study which I happened to see recently, there was terrific stability in virtual identity between the result in the open phase and the result a fairly short time later.

Dr. Pledger: I agree with Dr. Temple that the analysis of comparing means is not the only one that is based on the study of randomization. Certainly you can define criteria for individual patients and then compare groups on that basis. That analysis is based on the study of randomization. The p values are generated by the randomization that you did and the p values are valid. Once you do something else, then you are assuming models and the biological or statistical models, and then it is a matter of how strongly you proved them. Apparently, Dr. Temple believes strongly enough to convince him in a lot of cases of antiarrhythmic drugs. I don't think how strongly I believe biologica models are very important.

Dr. Chun: I think that the individual data and also the mean data will come up to the same results. Anyway I will do both individual patients and me data.

Dr. Lin: Once a computerized system is set up, are the case report forms stil necessary?

Dr. Ehrreich: I answered that in the presentation. Yes, at least at the present time, the hard copy will be necessary until some other agreeable system that is printing from the computer at the agency will be established.

Dr Lipicky: I don't think they are necessary.

Dr. Temple: I do.

Dr. Harrison: Why do you think so.

Dr. Temple: Sheer reactionaryism. I don't think you should leap into the unknown until you know exactly how it is going to work. As Dr. Ehrreich said there are people who are computer hostile or computer nervous and long before you get completely rid of things like hard copy and the things that we know physicians are trained to use, you have to have a lot of experience with it. don't mean that you could never do it, but there are all kinds of interesting questions. How do we document our own review process to ourselves if somebody has sat in front of the screen and made some manipulations and satisfied himse with something going on, what is he going to do? There is nothing that his ne level supervisor can read. These are little silly problems that are probably solvable, but need to be solved.

Dr. Lipicky: You can generate as much hard copy as you want from a computer. Just press the button and you get it. If you want 1000 copies, you can get 1000.

Dr. Temple: But you have to pick among it and choose it. For example right now, if you want to see what a reviewer did, you can get the volume you looked at, eyeball it yourself, look at certain things, follow them out and his review says what he did, so you have an idea of what went on. If the interaction has been a personal one between the computer and the reviewer, how are you going to look at that? How are you going to talk about it?

Dr. Lipicky: You call the same computer.

Dr. Temple: And what do you ask for?

Dr. Lipicky: The same thing the reviewer asked for.

Dr. Temple: Do the whole thing yourself?

Dr. Lipicky: But that is what you would do if you were leafing through.

Dr. Temple: No, first of all the reviewer has gone through a process that took him several months. He has looked at this, then he thought about this, and he makes a long detailed process. How is the supervisor going to know what that process was?

Dr. Lipicky: How can the supervisor know what that process was simply because there are 1000 volumes and case report forms sitting in the shelf?

Dr. Harrison: What we are talking about now is a problem that we all have when we deal with computer data. We have that as investigators, companies have that as sponsors, and the FDA has that too, so it is a very big philosophic question.

Dr. Bigger: I am just going to say that this interesting concept you are raising. I think it is also true that you don't want to leap into the unknown without a lot of experience, but there are models for example, AT&T's programmers workbench, where you have 1000 programmers working on a big project and you are trying to document what they are doing in terms of the project goals and schedule and the thing automatically will fill in the schedule and somebody in a supervisory role can see what steps were done when and exactly what was done and what was reviewed and go through that step, you might use something like that as a model for how the audit was done.

Dr. Temple: That is if you know exactly what you think the reviewer ought to do.

Dr. Bigger: You can look at exactly what he did, what he ought to do..

Dr. Temple: They are individualized projects. As things are now, every NDA is put together different because companies have different attitudes about the way they ought to do it. There are certain general outlines, but everyone is constructed differently. Different decisions about who to include, who not to include, different ways of telling you who has been included, and who has not, each one is quite individual. Maybe, you could set up some rules so that they were all the same and when one came in you would say, Aha, that is just another antiarrhythmic.

Dr. Temple: But it strikes me that we are talking about different things. Not having hard copy case report forms in existence right this instant, does not mean that one couldn't generate hard copy case report forms from the computer data base if one wanted them. I interpreted the question as if you have a computer system that has case report forms in the computer, whether in spite of that, it was absolutely mandatory that one had hard copy, my answer was you wouldn't need hard copy unless there was some special circumstance that required it.

Dr. Harrison: If it were formatted appropriately so that you could get it if you got the disk, and you could print them out if you wanted them. Not that there would be no case report forms and you could print them if you wanted it.

Dr. Lipicky: Not that there would be no case report forms. The case report forms would exist on magnetic media, one would be able to get whatever case report forms one wanted anytime and get all of them.

Dr. Temple: I probably delivered my diatribe on the wrong subject, but what my interest was is hard copy of various analyses, lists, tabulations, and all that stuff. The actual case reports are used very intermittently, and we are talking about different things.

Dr. Harrison: I think that as Tom Bigger pointed out there are examples in other groups other than medical, who have utilized ways of data on a computer being analyzed and being able to be followed and monitored very carefully and I think that can be worked out in this area too. It hasn't been worked out fully.

Dr. Morganroth: My question is to the FDA members here, is why haven't you as of this time insisted on at least some relatively specific guidelines for how NDA's on a type of agent like antihypertensive, antiarrhythmic etc. be submitted so that your supervisory process of the project officers can be more logical and probably more accurate and that the whole review process might not be greatly sped up. Why hasn't that been done?

Dr. Temple: Just take any antiarrhythmic as an example. The last oral antiarrhythmic was approved 6 to 7 years ago and the next one in a week or two. Between that time there was none. Who knows when the next one will be. The point is that the technology changes, the study designs change, the ways to measure everything change. It is very hard to say what the right way to do it is, when you see them only infrequently and when even the ones that we are going to get in the next few years have been developed over a period of 7 or 8 years. All the study designs are different, all the measurements are different. There are certain things you can say about general ways of presenting it, but an exact cookbook is very hard.

Dr. Morganroth: I agree with that of course. But, I am saying look at study tables. So many companies have them in different order and they have different rules for exclusion and inclusion of patients. I am not talking about how the study should be done or what the definition of efficacy should be or how they should be analyzed, I am talking about their format. EKG data is pretty much P-R, QRS, Q-T intervals etc. It does not vary so much.

Dr. Temple: For example, who to include in a study, who to exclude is a matter of enormous controversy. Biostatisticians have one view..

Dr. Harrison: That is true of the IRS too but you have guidelines about who and what should be included.

Dr. Temple: I don't think it is exactly the same.

Dr. Harrison: I have spent a lot of hours talking to sponsors about submitting an NDA, one that one person could walk in the room and carry in and put on the desk. I bet you one like that, if it were put togethr appropriately could be looked at and evaluated in a period of a few days or weeks, not five years. I believe there ought to be some way to do that, so one man could walk into your office and carry the NDA.

Dr. Temple: It isn't the quantity of data. In the area of arrhythmias it is the incredible study designs that are used, the kind of data collected, and things like that. Even though Dr. Ehrreich pointed out how long it would take, I have to admit, we do not review every page. You probably didn't know that. Reviews are targeted so in the case of reports they are going to be of interest, and certain kinds of spot checking, so the NDA that we review is a much smaller document then that 10,000 page monstrosity. It is what we call the summary of about 5-6000 pages. It is much more a matter of a candid guide to what is in there that we need and if someone could put that in a volume, it would make it easier. It isn't so much the mass of data that is the problem.

Dr. Lin: This morning we saw the differences in pharmacological response of antiarrhythmic agents secondary to differences in drug metabolism, which in turn could be attributed to the pharmacogenetic factors. Are there pharmacogenetic tests you can utilize before you submit a patient to a specific arrhythmic agent, such that you can predict which group this patient will fall in?. In other words, would it be subject to rapid metabolism or slow metabolism. For antiarrhythmics are there such tests? Have these been developed?

Dr. Harrison: I think you are asking a very broad based question and trying to figure out who metabolizes and who does not such as Dr. Woosley pointed out, this is a multiple question. There are multiple ways drugs are metabolized, there is not a single one, but there are markers, and if you know what you are looking for, there are markers for rapid and slow metabolizers. I don't think there is a general marker that can be broadly used. Can we move on to other things about automatic data analysis? One of the questions I had is whether you could have the data qued properly and people be able to get at various parts of it and where you could have an online analysis as you were going along say at the FDA, but for everytime a look is taken at the data, you pay a price.

Dr. Lipicky: I think you only pay a price if you make a decision on the basis of the look.

Dr. Harrison: But you are telling me you could do that without making that decision.

Dr. Bigger: This is an area of my concern also. I would phrase it a little differently and then we could discuss it a bit. Implicit in Dr. Ehhreich's discussion it seemed to me was when he kept talking of stopping a trial or being concerned about stopping one of the treatments in the trial, it was implicit that we are talking about severe proarrhythmic effects or actual deaths or something, because why would you care if the trial was just to look at antiarrhythmic activity and the too few patients were responding with any antiarrhythmic effect. So it had an implication as far as I was concerned, that had death as an end point is a very complex subject. It seems to me that most trials are done because of the relative rareness of death as an endpoint with a minimal power to begin with, usually set at 80%, so you only have an 80% chance of finding a differences between the treatment and the placebo if it exists. You get a 20% chance of not finding it even if it does exist. Then when you take part of the data, like say even if you have collected 1/4 of the number of patients that were computed for the sample size using 80% power and that is pretty uncomfortable when you think about it to begin with and because rare events tend to come in clusters at times and so forth then you look at it and you are uncomfortable with it, but what in the world can you do with that. In my own mind, in a mortality trial, I think you would be better off not to look at it. It is a very difficult question about when to look, and if you see a little cluster, and you know what can happen in sampling theory, what to do about it, what a terrible burden that you haven't had in the past.

Dr. Temple: Dr. Ehrreich described it as an experiment and it was an experiment to look at a lot of things. One is whether it could be done at all and another was what the consequences of it are, and we are very sensitive to the point you made. One person with the problems of the world on his shoulder trying to be a safety committee is something that I wouldn't like to take on. I think it is very hard.

Dr. Bigger: What I am getting at is if you are looking for something that you will only find say with an 80% chance when you have all the data, say 3000 patients, and you have 350 patients, what can you do with that.

Dr. Temple: In a mortality study we are in the same position that the safety committee is perhaps, except there is only one person, no moral arbiter, no minister, no community representative, no nothing. Just one person having to make a decision. We talked all this out before hand. One possibility is to say, that we won't make any kind of decision like that. We simply will not intervene. Whether that is practically do-able, is hard to say, because people feel moral obligations, even though they would like in principle not to intervene. You do pay a statistical price if you might have intervened. Your failure to intervene is a decision and it means that the study wasn't going to the wrong way and there is a certain amount of statistical bias in that at least potentially. However, it really is an experiment we really are going to think about it before we try to do it. The other thing probably is self-evident. We don't have anything like the resources to monitor any large fraction of all studies. That is just out of the question, because it is a lot of work.

Dr. Campbell: Can I just hear how you would approach the concept of doing a mortality study on a sequential statistical plan, in that mortality is not usually a debatable end point and lends itself to a statistical approach defining boundaries of risk and efficacy might not that get round the problem that you would pre-define those boundary zones and you would plot and chart on a sequential basis how patients are performing.

Dr. Temple: All I meant is that we would have to be part of the statistical analysis. We would have to be figured in and if we are watching it day-by-day, that is a larger number of looks, then even an ordinary safety committee looks at. There are ways of looking day-by-day, but the point is that we would have to be figured.

Dr. Campbell: You could pre-agree the boundary zones. In other words, you would set guide lines for these boundary zones and thereafter, you would be looking at it, but you wouldn't be an independent factor that might terminate the study prematurely.

Dr. Lipicky: Correct. In the experience of one, the basic agreement that was made was there were stopping rules in the particular study that we are looking at that was a mortality end point study. Those stopping rules were to be enforced. They were pre-agreed to and there was a safety committee that was responsible for determining whether those stopping rules were arrived at or not and that was what was to dictate the stopping or the not stopping of the study on the basis of some kind of statistical evidence that there was some kind of differnce. The facts are however, that what was observed, once the study started, was that there was a pattern of mortality that was occurring that wasn't anticipated prior to the study starting, which raised the question with respect to whether or not the study was being conducted in an appropriate way. It wasn't the data dependent question, it was, are the patients being cared for properly question. Then what do you do? I really do believe from the vantage point of stopping and starting, that is manageable, but in our experience of one, what wasn't manageable was you got into a sort of moral and ethical problem, a philosophical problem, that arose from examining the data. That was a significant problem because there were no rules for that ahead of time.

Dr. Bigger: I guess what I am getting across was that I was blown away by your presentation and the implications of it, because what is the relationship then between the regulatory agency and the traditional safety and data monitoring board that usually is eminent statisticians and clinical trial people who have been selected for their ability to judge the data and so forth, and you are in the loop too, and how do you interact with each other, with investigators, etc. It seems to me it will take years of debate or at least some months of debate to figure out how to process the data right on line.

Dr. Lipicky: What you are saying is true, however, in actual practice, as we learned from an experience of one, it is a tough problem. However, in concept, it is not. If there is a safety monitoring committee and there are stopping rules that in terms of deciding that something happened in the study with respect to the science of the study, that is pre-defined and clear. Whether or not there is a computer system online, and whether or not a government agency is monitoring that system, if adverse reactions occur that are unexpected, mortality isn't the problem, but everyone's lungs are developing something, or livers are rotting, or whatever it is, that generates 1639 or whatever that is called and the FDA still would have to say, hey we think you ought to stop the study because everybody's liver is disappearing. So having access to that data because it is online, doesn't alter that relationship one iota, O.K. So in concept, it is fairly easy I think, but in actual practice, it turns out one has to practice it a while to work it all out and actually have it operate.

Dr. Harrison: I think this is a very interesting subject. I can see the major advantage to sponsors with multicenter studies, having terminals on site and data collection in a central location seemed to me to be very cost effective now and a great deal of information come from that kind of activity and I think that is going to be something that we will see a lot of in the future.

Dr. Sami: Dr. Cohn, how do you analyze in your final analysis of mortality or arrhythmic mortality the guy who is driving a car on the highway, and then dies after the car accident. Was that an arrhythmic death?

Dr. Cohn: Those are the problems you have to deal with if in your best judgment, the car accident occurred as the result of the individual dying, then it fits in that category, if it doesn't, then you may have to censor it, and these judgmental things have to be carried out blindly with as much forethought as possible before you undertake the study.

Dr. Sami: I don't think the statistician would agree with you that you can use your own judgment. When you set the trial and design, you have to agree to the fact that total mortality is the only honest way of looking at comparing two groups of people and if you are going to sub-define your population, then I think you are really going to bias the study. You are going to start judging whether in your opinion, this patient who has had an accident, maybe was drunk that day or maybe he did have an arrhythmia.

Dr. Cohn: I think you are evading the real issue. Yes, it is difficult. It is much more easy to measure witnessed cardiovascular death. But if sudden death represents only 50% of the total cardiovascular mortality and you want to pick up a 20% reduction of sudden death, and you being with half the deaths which are not going to be impacted with your therapy, you are going to end up with only half the power of your study. I don't think that careful analysis of the deaths afterwards, blinded so that the analyzers are unaware of the therapy should be completely excluded as an approriate way to handle some of these questionable deaths.

Dr. Temple: I was just going to say that one should before deciding not to use total mortality, calculate just how much it would cost you to use total mortality, that is how much bigger your sample size would have to be because problems that are raised are formidable and will always create suspicion in at least some people that it is not as solid as it was. One of the things in people, for example, there are certain sets of people who are likely to die of the event you are interested in with a relatively high frequency and secondary prevention studie for example, almost deaths are cardiovascular deaths, so you don't lose anything much by counting total deaths. It doesn't cost you much. If you can define populations here, for example people with cardiomyopathies, in whom the possibility of ischemic death is somewhat less, you may really have the ideal population for carrying out a trial and just counting total mortality.

Dr. Cohn: Certainly I don't mean to infer that you don't count total mortality. Obviously all studies have to look at total mortality, but I think it is also appropriate to try to sub-set if you can, because you could have a therapy which impacts favorably on the arrhythmic deaths and has a negative impact on pump failure and you end up with no impact of the drug, and you might say this drug has had no effect whatsoever, where it might have had a rather profound effect on changing the distribution of deaths and if you don't set you criteria up to begin with you will lose that information.

IV. NEW ANTIARRHYTHMIC DEVICES

15

DVI VS DDD PACEMAKERS - PROARRHYTHMIC OR ANTIARRHYTHMIC?

S.R. SPIELMAN, M.D.

INTRODUCTION

Pacemaker utilization in the United States has increased at a remarkable rate over the last 15 years. At a time when battery replacements have diminished steadily with the development of lithium power sources, Parsonnet et. al. have still demonstrated a 600% growth rate in implants over the last 10 years(1). As devices have proliferated they have also become more technologically advanced and more complex with a significantly broader application to both brady and tachyarrhythmias. Although single chambered devices still account for the majority of U.S. implants, it has been estimated that dual-chambered pacemakers will ultimately be responsible for 25-40% of the pacemaker market(2). Moreover, the continued inadequacy of antiarrhythmic drug therapy in preventing life-threatening arrhythmias and the increased short-term salvage rate for patients who have suffered cardiac arrests both support a rapidly developing role for specialized pacemakers for antiarrhythmic use.

This chapter will discuss DVI and DDD pacemakers only. It is well-established that these devices are physiologically sensible and technically feasible to implant. As antiarrhythmic devices, however, they are somewhat limited; in fact, their complexity is such that they may themselves actually cause arrhythmias. We will therefore review the basic function, indications for implantation, antiarrhythmic application and proarrhythmic potential of both DVI and DDD pacemakers.

DVI PACEMAKERS

A DVI pacemaker paces in the atrium and ventricle (D), senses only in the ventricle (V) and has an inhibited (I) mode of response. These pacers may be either "committed" or "non-committed" depending on whether or not ventricular sensing continues during the AV interval. In a "committed" unit, once an atrial stimulus is emitted a ventricular stimulus must follow and one is likely to see frequent episodes of fusion and "pseudofusion" in the ventricle. In a "non-comitted" unit, because atrial sensing resumes during the AV interval, after a brief "blanking" period following an atrial stimulus, one may observe pacing of the atrium and ventricle, atrium only or total inhibition.

DVI pacemakers have proven hemodynamic benefit, especially in patients with 1:1 VA conduction during ventricular pacing who are particularly prone to "pacemaker syndrome" and in patients with carotid sinus hypersensitivity. In a large study by Stone et. al. in patients with sick sinus syndrome, DVI pacing offered significant symptomatic improvement over VVI pacing including relief of neurologic symptoms, angina, congestive heart failure etc.[3] DVI pacemakers are indicated in patients with sick sinus syndrome associated with AV nodal or His-Purkinje disease or in AV block with accompanying evidence of abnormal sinus node function where hemodynamic benefit from appropriately timed atrial systole can be expected.

DVI pacemakers clearly have arrhythmogenic potential because of the frequency with which atrial competition results from their normal operation. Atrial competitiion is a natural consequence of the inability of such pacemakers to sense atrially and the interaction between the VA interval and the patient's intrinsic rate. For example, if the base rate of the pacemaker is programmed to a rate of 80 (or 750msec) and the AV interval to 250msec, the resultant VA interval will be 500msec. This means that an atrial spike must occur 500msec after a sensed or paced QRS unless another ventricular depolarization appears in the meantime. This frequently results in atrial spikes falling in close relationship to naturally occurring P waves and may allow induction of atrial fibrillation (Figure 1). Furman and Cooper reported that 16% of patients with newly implanted DVI pacemakers developed atrial fibrillation that they had not had previously and suggested that pacemakers with atrial sensing were preferable(4). Other authors have not confirmed this high an incidence of induction of atrial fibrillation. In

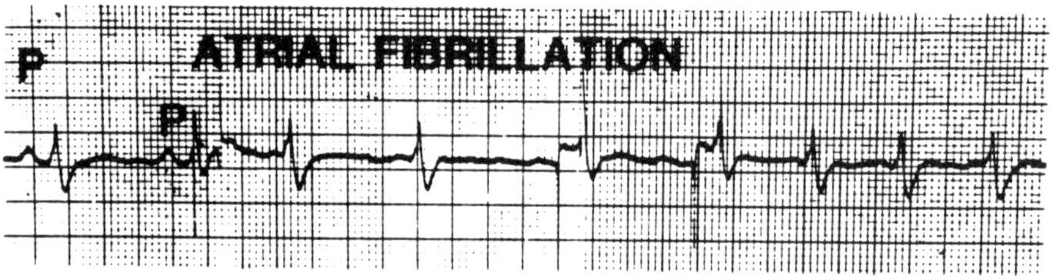

Figure 1 - Induction of atrial fibrillation by a DVI pacemaker. Lack of atrial sensing, inherent to such devices, results in induction of atrial fibrillation when a pacemaker stimulus falls in the atrial "vulnerable period". (Reproduced from Furman and Cooper: PACE 5:134, 1982, by permission of Futura Publishing Co., Inc.)

fact, in patients with brady-tachy syndrome in whom episodes of atrial fibrillation are bradycardia related, DVI pacers may actually decrease the frequency of episodes.

Recently Floro et. al. have proposed a modification of the DVI mode of pacing designed to solve the problem of atrial competition(5). They propose a DDI mode in which atrial sensing is present but atrial "tracking" is not. This is essentially AAI pacing with ventricular backup and is an interesting potential solution to the problem of atrial competition.

Several authors have attempted to apply the concept of dual site stimulation through DVI pacemakers to the prevention of AV reentrant tachyarrhythmias[6,7]. In patients with paroxysmal supraventricular tachycardia, AV pacing with a short AV interval may theoretically prevent arrhythmia recurrence by producing impulse collision in a region of slow conduction, thus precluding the initiation of reentrant beats. However, the adverse hemodynamics associated with AV pacing with a short AV delay are such that few of these devices have ever been implanted. Alternatively DVI pacing may be useful in a so-called "dual-demand" fashion to terminate an established SVT. A dual-demand device operates as a standard pacemaker to protect against bradycardia; during tachycardia it is automatically activated to pace at the same rate as the demand rate until the tachycardia terminates. The Medtronic 7007/8 is one such device currently under investigaton.

DDD PACEMAKERS

A DDD pacemaker can sense in both chambers (D), pace in both chambers (D) and is either triggered or inhibited or both (D). During normal function, when atrial activity falls below the base rate and atrial pacing is not followed by a ventricular event, this device would sequentially pace both atrium and ventricle with a suitable time delay. When atrial activity exceeds the base rate, the pacemaker switches into an atrial synchronous mode with ventricular pacing following atrial sensing provided the patient's own QRS does not occur during the programmed AV interval. Premature ventricular events inhibit both atrial and ventricular output and reset the pacemaker timing cycle. All current DDD units are non-committed and all employ some type of blocking mode to handle atrial arrhythmias that exceed the programmed upper rate of the pacemakers.

Multiple authors have demonstrated the hemodynamic advantages of "rate-responsive" pacing[(8)]. DDD pacemakers are indicated in patients with AV block and evidence of sick sinus syndrome, AV block with normal sinus node function and in patients with moderated sick sinus syndrome and AV nodal or His-Purkinje disease who maintain at least some ability to increase atrial rate with exercise.

DDD pacemakers may become arrhythmogenic as a natural consequence of their ability to sense in the atrium. The addition of atrial sensing essentially provides the electrophysiologic milieu necessary for the development of a reentrant tachycardia. This arrhythmia in pacemaker patients has been nicknamed "pacemaker-mediated" or "endless loop" tachycardia[(9)]. A DDD pacing system provides 2 antegrade routes to the ventricle, the pacemaker and the patient's own conduction system, but only one retrograde route to the atrium, the patient's own His-Purkinje system and AV node, since no conduction can occur retrogradely through the pacemaker. If a premature ventricular beat, for example, conducts retrogradely with a longer VA interval than the programmed atrial refractory period, this returning impulse will be sensed through the atrial lead and initiate a coupled ventricular response. If this ventricular event is then again conducted retrogradely a true reentrant tachycardia may ensue (Figure 2). In the last several years, the problem of pacemaker mediated tachycardia has largely been solved by the development of programmable atrial refractory periods. With VA intervals programmed to greater than 250 or 350msec, reentrant tachycardias rarely occur.

Another potential proarrhythmic consequence of atrial sensing and intact VA conduction is the development of ventricular arrhythmias. Freedman et. al. reported the initiation in a patient with a VDD pacemaker of sustained ventricular tachycardia following retrograde conduction of a ventricular paced beat which resulted in a closely coupled ventricular response[10]. Fortunately, induction of such life-threatening arrhythmias is quite rare.

By themselves, DDD pacemakers have no true anti-tachycardia efficacy. Several manufacturers have however tried to incorporate into standard DDD devices some anti-tachyarrhythmic functions. The Cordis Gemini, for instance, has a programmable triggered mode which allows manually

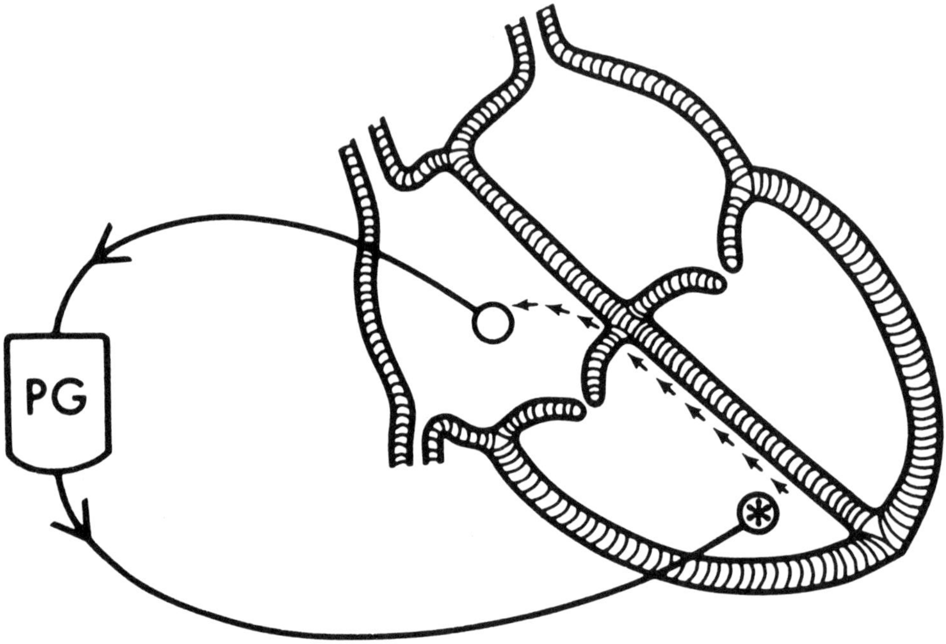

Figure 2 - Schematic diagram of the mechanism of pacemaker-mediated tachycardia. The open circle denotes sensing; and the circle with the asterisk, sensing and pacing. Each time the ventricle is depolarized by the pacer stimulus, the impulse is retrogradely conducted to the right atrium and detected by the atrial electrode. The pacemaker then triggers the next stimulus into the ventricle and cycle is repeated. (Reproduced from Tolentino et al: PACE 5:258, 1982, by permission of Futura Publishing Co., Inc.)

initiated burst pacing termination of supraventricular or ventricular tachycardia. More recently Medtronic has developed the Model 7007/8 DDD pacemaker which has programmable automatic anti-tachycardia modes including burst and dual-demand in addition to its standard bradycardic function (Figure 3).

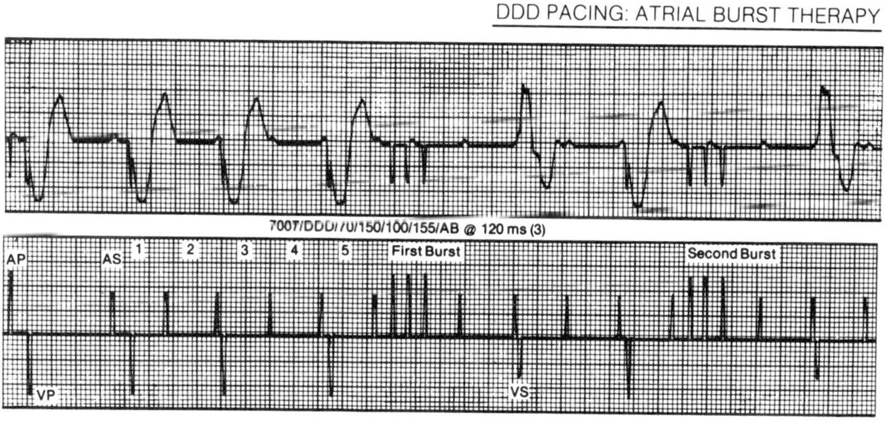

Figure 3 - An episode of atrial tachycardia occurs, is detected (five consecutive atrial events faster than the atrial upper rate) and is treated with a burst of three pulses. Since this first burst does not terminate the episode, a second burst is emitted, after the pacemaker counts five consecutive atrial events which are again faster than the atrial upper rate. (Reproduced from The Medtronic 7007/7008 Technical Manual by permission of Medtronic Corp.)

CONCLUSIONS

Advances in the design of DVI and DDD pacemakers in just the last 1-2 years have allowed application of dual chambered pacing to a steadily increasing number of patients with a steadily decreasing incidence of complications and proarrhythmia. As antitachycardia devices by themselves they are not sufficient but it is likely that more sophisticated antitachycardia devices of the future will incorporate dual chambered pacing capability into their ultimate design.

REFERENCES

1. Parsonnet V, Bernstein AD. Cardiac pacing in the 1980's: Treatment and Techniques in Transition. JACC 1:339-54, 1983.

2. Sutton R, Perrins J, Citron P. Physiologic cardiac pacing. PACE 3:207-19, 1980.

3. Stone JM, Bhakta RD, Lutgen J. Dual chamber sequential pacing management of sinus node dysfunction: Advantages over single-chamber pacing. Am Heart J. 104:1319-1327, 1982.

4. Furman S, Cooper JA. Atrial fibrillation during AV sequential pacing. PACE 5:133-35, 1982.

5. Floro J, Castellanet M, Florio J, Messenger J. DDI: A new mode for cardiac pacing. Clinical Progress In Pacing and Electrophysiology 2:255-260, 1984.

6. Castellanos A, Waxman HL, Moleiro F, Berkovits B, Sung RJ. Preliminary studies with an implantable multimodal AV pacemaker for reciprocating atrioventricular tachycardias. PACE 3:257-65, 1980.

7. Sung RJ, Styperek JL, Castellanos A. Complete abolition of the reentrant supraventricular tachycardia zone using a new modality of cardiac pacing with simultaneous atrioventricular stimulation. AJC. 45:72-78, 1980.

8. Kruse I, Arnman K, Conradson TB, Ryden L. A comparison of the acute and long-term hemodynamic effects of ventricular inhibited and atrial synchronous ventricular inhibited pacing. Circulation 65:846-855, 1982.

9. Tolentino AO, Javier RP, Byrd C, Samer P. Pacer-induced tachycardia associated with an atrial synchronous ventricular inhibited (ASVIP) pulse generator. PACE 5:251-59, 1982.

10. Freedman RA, Rothman MT, Mason JW. Recurrent ventricular tachycardia induced by an atrial synchronous ventricular inhibited-pacemaker. PACE 5:490-94, 1982.

16

PACING FOR VENTRICULAR TACHYCARDIA

JOHN D. FISHER, M.D.

INTRODUCTION

Pacing has been used for more than two decades in the treatment of ventricular tachycardia (VT) (1). There has been a steady evolution in the indications and applications of pacing in the therapy of VT. Within the medical community, the enthusiasm for VT pacing has waxed and waned, and implanted pacemakers for VT therapy continue to be used in a small minority of patients referred to specialized centers for treatment of these arrhythmias. Improved diagnostic and therapeutic abilities of implantable devices may greatly increase the number of patients suitable for therapy, but regulatory issues remain to be resolved.

Ideal Antiarrhythmic Therapy

Elimination of the tachycardia substrate would be desirable. Cardiac surgery can be highly efficacious, but is a major undertaking, involving substantial morbidity and mortality. Nonsurgical electrical catheter ablation or fulguration has been introduced recently, and may be effective in suitable patients. Drug therapy appears to be the most benign approach, in spite of the capacity of nearly every medication to aggravate arrhythmias in some patients. However, the "universal medication" has not yet been identified, and the effective drug or combination is usually identified only after long, tedious, and often gruelling drug trials, during an extended hospital stay.

Pacing would seem to have several advantages as a contender for ideal antiarrhythmic therapy. Pacemaker implantation is relatively routine minor surgery. Once implanted, the device is essentially inert until it is called upon to deal with an arrhythmia. In contrast to medications, there are few biologic or toxic side effects. The compliance problem is eased, because the patient is freed from the rigorous daily schedule of attention required with pharmacologic therapy. Negative aspects remain: Unlike other therapies, implanted devices have previously played little role in the prevention of arrhythmias. Automatic units have had difficulties distinguishing between physiologic sinus tachycardias and pathologic arrhythmias, and there has been the very real danger that therapeutic interventions by the pacemaker could actually result in acceleration of the VT or precipitation of ventricular fibrillation (1,6,7). These problems present both challenges and limitations to pacemaker therapy of VT. The technological problems are now on the verge of being solved. It remains to be seen if regulatory hurdles based on very legitimate concerns can be overcome.

Although we are presently concerned primarily with anti-VT pacing, it should be pointed out that many of the same techniques are applicable for supraventricular tachycardias (SVT). Except in certain patients with pre-excitation syndromes, the dangers of acceleration are much less than for anti-VT pacing. Several factors have thus far contributed to the limited use of pacers for SVT: these arrhythmias are often responsive to lower doses of medications than VT; responses to pacing may vary widely depending on neurohumoral tones; and there are regulatory limitations, based partly on concern that widespread availability of these units might lead to inappropriate use in patients with VT.

Pacing for Prevention of Tachycardia (1-4)

Ventricular pacing at moderate rates (60-100 bpm) is quite effective in preventing ventricular tachycardias occurring in the presence of bradycardias, especially heart block. VTs associated with long QT interval syndromes are readily suppressed by ventricular pacing. Pacing is much less effective in dealing with VT unrelated to bradycardia or a long QT interval. There have been occasional successes with somewhat more rapid ventricular or atrio-ventricular pacing at chronic rates between 100 and 120 bpm, but such pacing rates are often not well tolerated, and the overall incidence of success is not high.

More sophisticated techniques have been introduced to prevent VT. Among these are the variations on "orthorhythmic pacing" (8) in which the pacer recognizes various warning arrhythmias, and responds with stimulation in a pattern designed to "abort" the incipient VT. Also under investigation is the possibility of preventing VT by using electrophysiologic changes caused by stimulation during the refractory period. (9).

Termination of Tachycardia (1-8)

Magnetically activated implanted pacemakers have been used since the 1960s for slow, competitive, underdrive pacing for termination of VT. Few patients respond reliably to this approach, (7) and other methods were soon introduced. Multiple extrastimuli, and bursts of rapid ventricular pacing are more likely to achieve termination, but also result in a higher incidence of acceleration (7). During the 1970s and early 1980s, a host of pacing patterns were introduced in the hopes of achieving the best ratio between termination and acceleration (1). Patients clearly have benefitted from implanted anti-VT pacers, it appears at the present time. The problem of occasional acceleration still persists, however, limiting widespread applicability of the method.

In spite of these limitations, anti-VT pacemakers have been highly effective in treating selected and carefully evaluated patients. Our own series, (1) now amounting to about 25 patients, includes many whose lives appear to have been saved on several occasions by their anti-VT burst pacemakers. Other patients, unresponsive to medications and not suitable for surgery, have had countless episodes of VT terminated without the need for repeated hospitalization or external countershock. None of our patients have died under circumstances suggesting acceleration from their implanted units.

"Implantable Electrophysiologic Laboratories" (10)

Disease processes which produce VT are often progressive. Response to therapy may change over a period of months to years. Some patients with implanted anti-VT pacers require concomitant drugs to reduce the incidence of VT, lower its rate, or improve its responsiveness to anti-tachycardia pacing. Many currently available implantable pacemakers can be used to perform non-invasive programmed electrical stimulation (PES) so that responsiveness to antiarrhythmic drugs and anti-tachycardia pacing therapy can be followed on an outpatient basis. By reducing PES from a major catheterization laboratory procedure to an outpatient check-up, the long term management of the patient is greatly facilitated. Patient safety is also enhanced, since outpatient PES provides for periodic and often prophylactic changes in therapy.

Diagnostic Algorithms (1)

Manually activated implantable anti-VT pacemakers remain a viable although somewhat inconvenient option. Patients are generally instructed to go to the emergency room, where their VT can be terminated using the pacemaker under ECG control, with "rescue" available in the event of acceleration. Implicit in this type of unit is a major degree of inconvenience for the patient, as well as a VT which is well enough tolerated to permit the patient to make his way to the emergency room. Automatic units are clearly preferable, but their use has so far been limited by the problem of acceleration and limited diagnostic ability.

Automatic implantable burst pacemakers of the late 1970s and early 1980s, used a simple rate-duration algorithm for diagnosis of VT: a prescribed number of ventricular events at shorter than a prescribed cycle length was diagnosed as VT, and the anti-tachycardia pacing response was initiated. These units were incapable of differentiating between ventricular tachycardia, and other arrhythmias in which therapy would be contraindicated such as sinus tachycardia or atrial fibrillation with a rapid ventricular response. Since burst pacing is itself a reasonably effective method of initiating VT, it is possible that VT could be initiated by bursts delivered in response to sinus tachycardia, and that this pacer-induced VT was then terminated by subsequent bursts. These units had marginal memory and telemetry capabilities, making it difficult to substantiate patient reports of antitachycardia interventions by the implanted unit.

Implanted pacemakers are now being introduced which incorporate elaborate algorithms for differentiating between types of tachycardias. The units are highly programmable, allowing a major degree of tailoring to individual needs. Most algorithms are presently based on single chamber sensing, and it is likely that dual site or double chamber sensing will further improve the accuracy of the algorithms. The memory and telemetry capabilities of these units are such as to make them virtually miniature Holter monitors, so that the physician can document the efficacy of therapy, and make adjustments as needed. In spite of these improvements in diagnostics, the therapeutic problem of acceleration remains. Here too, a solution may be in sight.

Dealing with Acceleration: The "Big Bang Back-Up"

The automatic implantable defibrillator has proved able to revert the most malignant of all arrhythmias back to normal sinus rhythm (11). The defibrillator is a large unit, requires some form of thoracotomy, and discharges are somewhat uncomfortable for the patient, especially if frequent. Tests of transvenous cardioverting systems have had mixed results: low energy shocks are comparable to pacing techniques in safety and effectiveness, but are much less comfortable (12,13). High energy transvenous shocks also appear to be capable of terminating ventricular fibrillation (13). The combination of sophisticated pacing algorithms and back-up cardioversion/defibrillation capability would substantially expand the role of implanted antitachycardia devices. The potential of this approach would be further improved if the size of the device and its cost were reduced, and the need for thoracotomy avoided. Devices with these characteristics are presently under development by several manufacturers.

REGULATORY CONCERNS

There are several legitimate issues, which combine to form a major impediment to the more widespread use of antitachycardia devices.

1. Implantation of an antitachycardia device must almost invariably be preceded by a detailed electrophysiologic evaluation. This pre-implant evaluation is necessary in patients with SVT and pre-excitation syndromes to be sure that atrial fibrillation will not result in an extremely rapid ventricular response, and to determine the best lead position and antitachycardia pacing pattern. For patients with VT, electrophysiologic testing is required to determine the best pacing technique, and to estimate the likelihood or incidence of acceleration. Many VT patients have multiple VT morphologies, and each of these must be evaluated. For both VT and SVT, the contribution of concomitant drug therapy to successful pacer therapy must be evaluated.

2. Detailed electrophysiologic testing requires highly trained and experienced personnel. These are not available at most hospitals.

3. The federal government, through the United States Food and Drug Administration (F.D.A.) is responsible for ensuring and evaluating the safety and efficacy of therapeutic modalities.

4. Once a device is commercially available, there is no way to restrict sales exclusively to institutions with the capability of performing necessary pre-implant electrophysiologic or other testing. Thus there is concern that inappropriate and possibly dangerous implants may take place.

5. Patients who feel that they have been injured by inappropriate antitachycardia devices are likely to take legal action against their

physician, the hospital, the device manufacturer, and possibly regulatory agencies.

These concerns have led to a virtual paralysis in the introduction of antitachycardia devices. Devices developed in the United States after careful thought and consultation with leaders in the arrhythmia field, are now being implanted overseas months to years before they are available in the United States even on a research basis. Several companies have transferred their main research and development efforts in the antitachycardia field overseas, probably to the detriment of this country's scientific, technological, and medical progress. The costs of carrying out complex clinical trials are extremely high, especially considering that the market for antitachycardia devices remains theoretical and potential. The process of introducing new antitachycardia devices is therefore slowed by the complex regulatory process, uncertainty as to the market, high research, development, and clinical trial costs, and anxiety on all fronts regarding potential lawsuits.

A case in point is the saga of the Intermedics Cybertach pacemaker. This is a multiprogrammable automatic implantable antitachycardia pacemaker, which began clinical trials in 1979. The premarket approval application (PMAA) was submitted in 1982. An F.D.A. Panel meeting reviewed the data in July, 1982. In the experimental protocol, Cybertachs had been used to treat VT in some patients, but the PMAA sought approval only for use in SVT. A concern was expressed by the F.D.A. Panel, by the manufacturers, and by the physician-investigators present regarding the use of the Cybertach for VT in the event that approval was granted for use in SVT. Although the Panel report appeared tentatively favorable, certain additional information was required, and conditions were proposed. In the absence, however, of effective methods of regulating the device after commercial introduction, both the F.D.A. and the manufacturer have been hesitant in trying to get the Cybertach onto the market, and in late 1984, the Cybertach remains in limbo. Device technology is progressing rapidly, and the Cybertach is already obsolescent. In view of the experience with this device, manufacturers will hesitate to invest the large sums required to develop more advanced models.

The key, therefore, is to devise a method of regulating the use of potentially dangerous devices after commercial introduction. Possible methods include special licensing by the federal government, certification of physicians by specialty boards or professional societies and a major focus on the medical/legal implications of working with antitachycardia devices. Since most devices are sold to institutions rather than individual physicians, there may be a role for institutions in deciding which physicians are authorized to deal with antitachycardia devices. Manufacturers might demand a waiver of liability (for whatever this would be worth) if the institution could not provide assurances of the qualifications of its physicians. Whatever the method of regulation ultimately adopted, it should come as soon as possible, with major input from government, manufacturers, physicians and professional societies.

CONCLUSIONS

The technologic barriers for widespread application of anti-VT pacing are now being breached. The major problem remains the absence of suitable controls on the use of potentially dangerous devices after commercial introduction. The establishment of appropriate controls, acceptable to government, manufacturers, and physicians, should receive immediate attention from all concerned, for the ultimate benefit of American patients.

REFERENCES

(Note: this is a selected list of references. Several are review articles or similar, containing more lengthy lists of references for the interested reader.)

1. Fisher JD, Kim SG, Furman S, Matos JA: Role of implantable pacemakers in control of recurrent ventricular tachycardia. Amer J Cardiol 49:194-206, 1982.
2. Fisher JD, Cohen HL, Mehra R, Altschuler H, Escher DJW, Furman S: Cardiac pacing and pacemakers II. Serial electrophysiologic-pharmacologic testing for control of recurrent tachyarrhythmias. Amer Heart J 93:658-668, 1977.
3. Heiman DF, Helwig J, Jr: Suppression of ventricular arrhythmias by transvenous intracardiac pacing. JAMA 195:172-175, 1966.
4. Lew HT, Match HW: Control of recurrent ventricular fibrillation by transvenous pacing in the absence of heart block. Amer Heart J 73:794-797, 1967.
5. Fisher JD, Kim SG, Matos JA, Waspe LE: Pacing for tachycardias: Clinical translations. PACE, 1984. In Press.
6. Fisher JD, Mehra R, Furman S: Termination of ventricular tachycardia with bursts of ventricular pacing. Amer J Cardiol 41: 94-102, 1968.
7. Fisher JD, Kim SG, Matos JA, Ostrow E: Comparative effectiveness of pacing techniques for termination of well-tolerated sustained ventricular tachycardia. PACE 6:915-922, 1983.
8. Gurtner HP, Gertsch M, Zacouto F: Orthorhythmischer Herzschrittmachner und salvenformige Herstimulation. Schweiz Med Wschr 105:33-38, 1975.
9. Prystowsky EN, Zipes DP: Inhibition in the human heart. Circulation 68:707-713, 1983.
10. Fisher JD, Furman S, Kim SG, Matos JA, Waspe LE: DDD/DDT Pacemakers in the treatment of ventricular tachycardia. PACE 7:173-178, 1984.
11. Mirowski M, Reid PR, Mower MM: Termination of malignant ventricular arrhythmias with an implanted automatic defibrillator in human beings. New Engl J Med 303:322-324, 1980.
12. Zipes DP, Jackman WM, Heger JJ, Chilson DA, Browne KF, Naccarelli GV, Rahilly GT, Prystowsky EN: Clinical transvenous cardioversion of recurrent life-threatening ventricular tachyarrhythmias: Low energy synchronized cardioversion of ventricular fibrillation in patients using a catheter electrode. Amer Heart J 103:789-794, 1982.
13. Waspe LE, Kim SG, Matos JA, Fisher JD: Role of a catheter lead system for transvenous countershock and pacing during electrophysiologic tests: An assessment of the usefulness of catheter shocks for terminating ventricular tachyarrhythmias. Amer J Cardiol 52:477-484, 1983.

17

TERMINATION OF VENTRICULAR TACHYCARDIA BY TRANSVENOUS CARDIOVERSION

ERIC N. PRYSTOWSKY, WILLIAM M. MILES, JAMES J. HEGER, DOUGLAS P. ZIPES

INTRODUCTION

Transthoracic shocks using relatively low energy, for example, less than 40 joules, often can terminate organized sustained ventricular tachycardia. However, transthoracic shocks with these energy levels are painful and not tolerated by the awake patient. Previous investigations by Mirowski et al (1) demonstrated that ventricular fibrillation could be terminated in humans with an implantable defibrillator. We hypothesized that shocks applied directly to the myocardium via a catheter electrode might terminate sustained ventricular tachycardia with even lower energy levels than those used transthoracically or those necessary for internal defibrillation. These well-tolerated shocks would enable cardioversion in the awake alert patient. This report details the development of such a system.

Transvenous cardioversion - temporary application

Initial studies (2) performed in open chest anesthetized dogs following recovery from myocardial infarction demonstrated that a synchronized shock of 0.008 to 1.0 joules delivered via a catheter electrode successfully terminated 83% of episodes of ventricular tachycardia with cycle lengths greater than 200 msec. Subsequent studies in humans showed that a similar catheter electrode system could terminate ventricular tachycardia, ventricular fibrillation (with higher energy settings), and atrial fibrillation (3-8). The cardioverting catheter electrode is inserted into a peripheral vein, we prefer the internal jugular vein although subclavian and antecubital sites have been used, and positioned under fluoroscopic guidance to the right ventricular apex. The catheter has two bipolar pairs of stainless steel electrodes that have a surface area of 2.5 cm^2/pair, and separation between each bipolar pair is 5 mm. The distal electrodes are located at the catheter tip and are separated from the proximal electrodes by 100 to 150 mm. The variable electrode spacing enables positioning of the proximal electrodes near the superior vena canal - right atrial junction in patients with different heart sizes.

To ensure appropriate catheter placement, measurements of pacing thresholds and sensing thresholds and slew-rate are obtained prior to connecting the electrode lead to a temporary cardioverter. The cardioverter can deliver shocks that have a truncated exponential wave form with tilts of either 27, 63 or 82 percent at various energy levels yielding shocks of 0.03 to 28 joules. After ventricular tachycardia is induced by programmed ventricular stimulation (9), the ability of the temporary device to sense and terminate tachycardia can be tested (3-5).

Recently, a multicenter trial investigated the efficacy of synchronized transvenous cardioversion in patients with recurrent ventricular tachycardia (7). Seventy-two of 96 patients (75%) who had episodes of ventricular tachycardia had at least one successful cardioversion with shocks delivered over the lead. Of 467 episodes of sustained ventricular tachycardia, 397 (85%) were terminated with restoration of sinus rhythm. Forty episodes (8.6%) showed no change in ventricular rhythm while 24 (5%) had ventricular flutter or fibrillation produced, and 6 (1.4%) had transient acceleration of the ventricular tachycardia. The mean cycle length of all ventricular tachycardia was 365 msec (range 170-620 msec), while those tachycardias successfully terminated had a mean cycle length of 375 msec (range 170-575 msec) and those unsuccessfully terminated had a mean cycle length of 318 msec (range 180-620 msec). The mean energy level of all shocks delivered for ventricular tachycardia was 1.5 joules with a range of 0.03-28J. Those shocks that successfully terminated ventricular tachycardia had a mean energy of 1.44 joules with a range of 0.03-28J. Unsuccessful shocks had a mean energy (of final shock) equaling 5.2 joules (range 0.03-28, S.D. 8.5).

Accurate synchronization of the shock within the QRS complex is essential. On occasion, even accurately timed shocks delivered within the QRS complex may accelerate the ventricular tachycardia to ventricular flutter or ventricular fibrillation (4). Another potential complication is the precipitation of atrial tachyarrhythmias that may result when the shock, synchronized to the QRS complex, falls in the vulnerable period or T wave of the atria. This occurred in 18 patients in the multicenter trial (7).

Transvenous cardioversion - permanent implant

The implantable unit (Medtronic, Inc.) weighs 95 g and is 57 X 73 X 97 mm in size (Figure 1) (8). It delivers shocks of programmable energies between 0.06-2.0 joules either manually, that is, triggered by an external programmer or automatically upon detecting tachycardia. When set in the automatic mode, the unit delivers up to 3 shocks at the initial energy setting, and if ventricular tachycardia continues, 2 shocks at 2.0 joules. The tachycardia detection algorithm includes programmable cycle lengths from 205 to 545 msec and consecutive intervals (at the tachycardia cycle length) programmable from 2 to 32. Therefore, for a rhythm to be identified as tachycardia by the device, it must have a cycle length shorter than the programmed tachycardia cycle length and be present at this cycle length for a fixed number of intervals. An optional criterion includes the absolute change in cycle length from the last normal cycle to the tachycardia cycle. This criterion is based on the assumption that the change from normal sinus rhythm to tachycardia is abrupt. Therefore, the device can be programmed not to recognize tachycardia unless the onset to tachycardia is accompanied by a critical shortening of cycle length (programmable between 0-200 msec). The implantable cardioverter also functions as a demand pacemaker with programmable rates between 30 and 100 beats/min. In addition, on a temporary basis, the device can be used for rapid asynchronous pacing at rates of 30-400 beats/min to initiate or terminate tachycardia. Finally, the device is capable of noninvasive programmed electrical stimulation, delivering a series of basic stimuli (S_1) and one or more premature stimuli (S_2, S_3, S_N) at programmable intervals. Sensitivity and output of the pacing stimuli are programmable. The unit also is capable of telemetering its settings, an electrogram, marker channel and battery status.

Eleven patients have received permanent implants at Indiana University. Prior to implant, all patients had multiple episodes of ventricular tachycardia induced by programmed ventricular stimulation and terminated with the temporary transvenous cardioverter system (see above). No patient had acceleration of ventricular tachycardia with cardioversion, but after implantation, one patient had four episodes of acceleration of the ventricular tachycardia following the shock, and thus is not a candidate for programming into an automatic mode.

Figure 2 illustrates the initiation of sustained ventricular tachycardia and its termination by a cardioversion shock. In the top panel, the first QRS complex is the last spontaneous sinus beat followed by eight paced ventricular complexes (S_1) and a premature complex (S_2). The S_2 occurs in the ventricular refractory period but the train of S_1 stimuli initiates ventricular tachycardia (last 2 beats, top panel). In the midportion of the middle panel, the command "initiate cardioversion" is delivered to the device, which then senses the programmed number of cycles, six in this instance. Then, the device charges to 0.2 joules and delivers a synchronous shock within the QRS complex (bottom panel) that terminates the ventricular tachycardia. Note the marker channel initially delivers negative square wave pulses indicating that it senses normal ventricular activation. Following the command, "initiate cardioversion", QRS complexes are then recognized as tachycardia (positive square waves). Note the larger positive square wave indicating delivery of a cardioversion shock (arrow).

After implant, one patient had eight episodes of sustained ventricular tachycardia terminated without incident by a physician activating the cardioverter. The device then was set in the automatic mode and successfully terminated five episodes of recurrent sustained ventricular tachycardia also without incident. This patient did very well, until on one occasion, the cardioversion shock terminated the ventricular tachycardia, but produced atrial fibrillation. When the rapid ventricular response during atrial fibrillation met the tachycardia detection algorithm criteria, repeated shocks synchronized to the QRS complex were delivered. In addition, during this period of time, the patient had episodes of spontaneous ventricular tachycardia detected and cardioverted. After approximately one hour of ventricular rates 150-180/min and repeated cardioversion attempts, a shock was triggered late in the ST segment and produced ventricular flutter that degenerated to ventricular fibrillation. The patient was resuscitated without any problems or residual CNS deficits. The event responsible for triggering the late

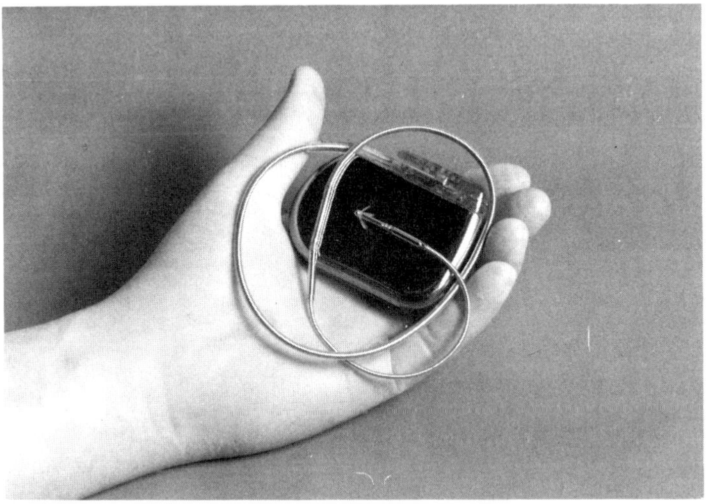

FIGURE 1: The cardioverter and tined lead held in the surgeon's gloved hand.

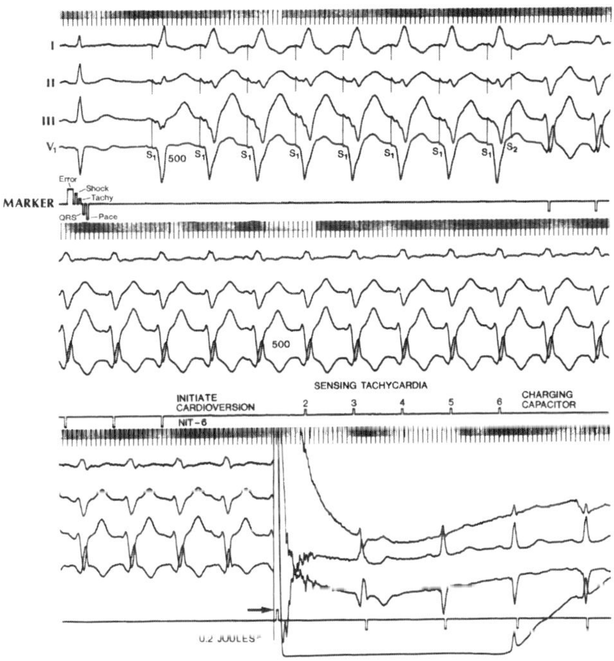

FIGURE 2: Ventricular tachycardia initiation and termination. From top to bottom in each panel the tracings are ECG leads I, II, III, V_1 and a marlar channel. Time lines = 50 msec for major division. See text for details.

shock was not clear. Because of this complication, the unit was programmed back to a manual mode. However, over time the patient's myocardial function deteriorated and ventricular tachycardia recurred frequently. Rather than returning to his physician's office for manual activation of the unit to terminate each episode of tachycardia, the patient requested he be reprogrammed to the automatic mode, fully aware of the potential risks of acceleration of ventricular tachycardia with transvenous cardioversion. In the automatic mode the patient once again had successful cardioversions of ventricular tachycardia, but died suddenly soon thereafter.

CONCLUSION

The data suggest that the development of this first generation device is useful to treat patients with recurrent sustained monomorphic ventricular tachycardia. The implantable cardioverter is particularly useful in patients who have concomitant bradyarrhythmias and require intermittent ventricular pacing. The small size, implantation by a transvenous route and capability of performing programmed electrophysiologic studies noninvasively are all major advantages. However, without backup defibrillation, the automatic mode must be used cautiously and very selectively.

ACKNOWLEDGEMENT

The authors appreciate the assistance of Elizabeth Darling, R.N., and Sue Faust, R.N. and the many Electrophysiology research fellows who participated in these studies, and Nancy S. Lineback for her secretarial assistance in the preparation of this manuscript.

Supported in part by the Herman C. Krannert Fund, Indianapolis, Indiana; by Grants HL-06308 and HL-07182 from the National Heart, Lung and Blood Institute of the National Institutes of Health, Bethesda, Maryland; and by the Attorney General of Indiana Public Health Trust and by the Roudebush Veterans Administration Medical Center, Indianapolis, Indiana; and by a Grant-In-Aid from the American Heart Association, Indiana Affiliate, Inc., Indianapolis, Indiana.

REFERENCES

1. Mirowski JF, Reid PR, Mower MM, Watkins L, Gott VL, Schauble JF, Langer A, Heilman MF, Kolenik SA, Fischell RE, Weisfeldt ML: Termination of malignant ventricular arrhythmias with an implanted automatic defibrillator in human beings. N Engl J Med 303:322-324, 1980.
2. Jackman WM, Zipes DP: Low-energy synchronous cardioversion of ventricular tachycardia using a catheter electrode in a canine model of subacute myocardial infarction. Circulation 66:187-195, 1982.
3. Zipes DP, Jackman WM, Heger JJ, Chilson DA, Browne KF, Naccarelli GV, Rahilly GT Jr, Prystowsky EN: Clinical transvenous cardioversion of recurrent life-threatening ventricular tachyarrhythmias: low energy synchronized cardioversion of ventricular tachycardia and termination of ventricular fibrillation in patients using a catheter electrode. Am Heart J 103:789-794, 1982.
4. Zipes DP, Prystowsky EN, Browne KF, Chilson DA, Heger JJ: Additional observations on transvenous cardioversion of recurrent ventricular tachycardia. Am Heart J 104:163-164, 1982.
5. Prystowsky EN, Heger JJ, Zipes DP: Transvenous cardioversion to treat tachyarrhythmias. In: J Morganroth and EN Moore (eds) Sudden Cardiac Death and Congestive Heart Failure: Diagnosis and Treatment. Martinus Nijhoff, The Hague, 1983, pp 185-198.
6. Yee R, Zipes DP, Gulamhusein S, Kallok MJ, Klein GJ: Low energy countershock using an intravascular catheter in an acute care setting. Am J Cardiol 50:1124-1129, 1982.
7. Kallok MJ, Fisher JD, Fletcher RD, Hartzler GO, Kehoe RF, Klein GJ, Prystowsky EN, Ruffy R, Zipes DP: Intracavitary cardioversion and defibrillation -- a multicenter study. (Abstract) Circulation 68(III):89, 1983.
8. Zipes DP, Heger JJ, Miles WM, Mahomed Y, Brown JW, Spielman SR, Prystowsky EN: Early experience with the implantable cardioverter. N Engl J Med 311:485-490, 1984.
9. Prystowsky EN, Heger JJ, Lloyd EA, Zipes DP: Clinical electrophysiology of ventricular tachycardia. In: DP Zipes (ed) Cardiac Arrhythmias. W.B. Saunders Co., Philadelphia, 1983, pp 253-273.

Address correspondence to:

Eric N. Prystowsky, M.D.

Krannert Institute of Cardiology

1001 West Tenth Street

Indianapolis, Indiana 46202

18

THE AUTOMATIC IMPLANTABLE CARDIOVERTER/DEFIBRILLATOR

Roger A. Winkle, M.D.
Debra S. Echt, M.D.
R. Hardwin Mead, M.D.
Vincent Gaudiani, M.D.
Edward B. Stinson, M.D.
Paula Schmidt, R.N.

BACKGROUND

It is currently recognized that patients with a history of recurrent sustained ventricular tachycardia or with ventricular fibrillation not associated with acute myocardial infarction or other reversible cause are at high risk for sudden cardiac death. Although some deaths in these patient populations are due to progressive congestive heart failure, approximately 2/3 are sudden and presumably preventable. Electrophysiologic testing has indicated that those patients treated with drugs which permit continued induction of sustained ventricular tachyarrhythmias have a 25-30% one year sudden death rate (1). Even patients treated with drugs rendering them noninducible at electrophysiologic study have a one year sudden death rate approaching 10%. Those patients who fail a number of drugs at serial electrophysiologic testing are considered excellent candidates for alternative therapies such as amiodarone, endocardial mapping with resection or cryoablation and/or automatic cardioverter/defibrillator (AICD) implantation. Long-term amiodarone therapy is associated with a high failure rate due to side effects, arrhythmia recurrences and sudden death. In our experience, the one year actuarial sudden death rate for patients receiving amiodarone is 16.7% and after 2 years only approximately 1/3 of patients remain effectively treated with the drug (2). Endocardial mapping and resection can be very effective in properly selected patients. However, many patients have poor LV function which results in a high operative mortality rate or have disorganized rhythms such as ventricular fibrillation or polymorphic ventricular tachycardia, or have multiple ventricular tachycardia morphologies which make endocardial mapping difficult. The AICD has been developed by Dr. Michel Mirowski and his colleagues over the past 15 years (3-9). Nearly 5 years of clinical experience with this device in man indicates that it can be implanted in patients with poor LV function with an acceptable risk. In our experience the operative mortality is approximately 2% and long-term follow-up indicates a 1 and 2 year sudden death rate of 2-5% (10). This manuscript describes the AICD device and associated hardware, the patient selection process to identify candidates who will benefit from the device, implant techniques (11) and the future role of the automatic implantable cardioverter/defibrillator.

HARDWARE

The AICD device we have most frequently implanted is the model AID-BR(R). This device is a rectangular shaped generator with dimensions of 11.2 cm x 7.1 cm x 2.5 cm with a Titanium case weighing 292 gm and an epoxy header to receive the 4 input leads. It utilizes local bipolar electrogram sensing from either a transvenous bipolar sensing catheter or a pair of standard epicardial screw-in leads. Shocks are delivered via a superior vena caval spring electrode which serves as the anode and a 10 sq cm surface area rectangular electrode which is sutured near the apex of the left ventricle and serves as the cathode. Some patients require 2 patch configuration with the shocks being given between 2 stan dard patches or 2 large patches each with 18-20 sq cm of surfac area. The patches consist of a Titanium mesh with a silasti backing. In the AID-BR(R) devices sensing is based on average heart rate derived from the local bipolar sensing electrogram (12). Any time a patient's heart rate exceeds the factory prede termined cutoff for approximately 5 seconds the device begins charging cycle and will deliver a shock approximately 8 second later (13-15 seconds after the onset of the tachyarrhythmia) Most units deliver 25 joules as a truncated exponential wavefor with a 60% tilt. If the initial shock is unsuccessful for termi nating the ventricular tachyarrhythmia, up to 3 subsequent shock of 30 joules each will be delivered. A "high energy" unit i available which delivers 30 joules for all 4 shocks. The devic has battery life for approximately 2-3 years of monitoring i relatively few shocks are given to the patient or has the capa bility to deliver greater than 100 shocks. If sinus rhythm ha not been restored after a sequence of 4 shocks, the device become inactive. The device can be activated and inactivated by applica tion of a magnet over the device for 30 seconds. If it is in th active mode, application of the magnet results in a audibl beeping tone synchronous with each QRS complex. After 30 second the tone becomes continuous indicating a change to the inactiv state. Application of a magnet during the inactive state result in the continuous audible tone and after 30 seconds the device i switched to the active state with a few seconds of random tone while the electronics are "warming up" followed by the synchronous QRS beeping. Transient application of the magnet in the active state results in a "magnet test" which initiates an internal charge of the device which is dumped through an internal resistor. By using an AID-CHECK(R) device the charge time of the unit can be monitored. Prolonging charging times indicate end of life. The device also provides a digital readout during each magnet test of the total number of shocks the patient has received.

PATIENT SELECTION

Patients receiving this device should have documented episodes of life threatening recurrent sustained ventricular tachycardia or ventricular fibrillation. Patients should not have hypokalemia, acute myocardial infarction or drug toxicity as a precipitating factor for their malignant ventricular tachyarrhythmias. Patients should undergo complete hemodynamic and angiographic evaluation including coronary arteriography and left ventricular angiography. Patients should have full intracardiac electrophysiologic study with attempted ventricular tachyarrhythmia induction. Patients experiencing multiple cardiac arrest with no ventricular tachyarrhythmia inducible at EP testing or those with a single arrest and having another indication for cardiovascular surgery may be candidates to receive this device without extensive drug testing. However, the bulk of patients receiving the device have arrhythmias inducible at electrophysiologic study and have failed a multitude of standard and investigational antiarrhythmic drugs. Patients should undergo continuous in-hospital monitoring in order to detect frequent supraventricular tachycardias or nonsustained ventricular tachycardias which would result in frequent triggering of the AICD. In our experience virtually all such patients can have their nonsustained arrhythmias adequately controlled with antiarrhythmic drugs so that these rhythms rarely represent a contraindication to placement of the device. Patients should also undergo maximal exercise treadmill tests in order to determine peak sinus heart rate. This rate can be compared to the rate of spontaneous and induced ventricular tachycardias to aid in rate cutoff selection for the implanted defibrillator. Patients must undergo considerable education concerning the device itself, the reason for implantation, the expected long-term follow-up and the discomfort related to receiving shocks. Because the device is currently investigational, patients must sign written informed consent prior to implantation. In general, poor left ventricular function rarely results in a contraindication to AICD implantation. Our criteria for adequate left ventricular function to undergo placement of a device is that the patient be ambulatory and off of all intravenous inotropic or vasodilator support.

IMPLANT TECHNIQUE

We have preferred to utilize a left thoracotomy incision as our primary approach for defibrillator patch lead placement in patients not undergoing other cardiac procedures requiring a median sternotomy incision. Other centers utilized a subcostal approach, a subxyphoid approach, or have utilized a median sternotomy as the primary approach even when other cardiac surgery is not contemplated. The patch lead is placed extrapericardially near the cardiac apex and 2 epicardial screw-in leads are placed in close proximity into healthy appearing left ventricular myocardium. The superior vena cava spring lead is placed percutaneously via the left subclavian vein using Seldinger technique through a 14 F sheath. These leads are tunneled subcutaneously to a left abdominal pocket.

Prior to placement of the AICD generator each patient undergoes extensive defibrillation threshold testing (13). Ventricular tachyarrhythmias are repeatedly induced in order to determine that 25 joules reliably terminates ventricular tachycardia and/or fibrillation and that there is a reasonable margin of safety (usually 10 joules or greater). Since ventricular fibrillation can be more difficult to terminate than ventricular tachycardia, we prefer to induce this arrhythmia in all patients even if the clinical rhythm has been ventricular tachycardia. Patients with unsatisfactory defibrillation thresholds have 2 large patch leads placed through the thoracotomy. One lead is generally placed over the posterolateral left ventricle and the other lead placed anteriorly over the right ventricle. Approximately 15-20% of patients require this lead configuration and it works in more than 2/3 of these difficult cases to provide satisfactory defibrillation thresholds (14). Once satisfactory defibrillation thresholds have been obtained the generator is connected to the leads, the lead caps are placed and all incisions are closed. We generally leave the device in the inactive state for the first few days after surgery and to treat any arrhythmias in a standard manner. The patient should undergo continuous 24 hour monitoring during the first day on which the device is activated. Each patient is restarted on antiarrhythmic therapy (we discontinue myocardial depressant drugs preoperatively whenever possible) and returns the electrophysiology laboratory for a predischarge testing of t device. The purpose of this test is to make certain that the AI functions properly with the patient in an nonanesthetized stat to assess the patients level of consciousness at the time defibrillator discharge and to permit the patient to experience shock prior to hospital discharge. All patients are trained the technique of cough CPR and instructed to begin coughing shou they experience symptoms of palpitations or light-headedness.

Patients requiring permanent pacemakers in addition to t AICD must have careful attention paid to the pacemaker type to certain that the pacer stimulus is not sensed by the AICD. A permanently implanted pacemakers must be bipolar and whenev possible the pacemaker lead placed at considerable distance fr the AICD bipolar sensing lead. Pacemakers with programmable ou put amplitude are preferred.

FOLLOW-UP

Patients are followed at 2 month intervals. At these visi magnet tests are performed in order to determine the number shocks the patient has received and to assess battery statu Patients receiving infrequent defibrillator shocks are advis only to call their physician and are not required to seek medic attention. Patients receiving several shocks in a matter minutes are advised to summon the paramedics. Patients receivi several shocks in a matter of a few days are placed on continuo magnet tape ambulatory ECG recordings in order to document t cause of their shocks. We have had considerable difficulty assessing the arrhythmias causing shocks based on clinical grounds alone. Shocks are preceived by the patient as being unpleasant but not necessarily painful. On rare occasions patients have received large numbers of shocks over a short period of time and have had some transient psychological disturbance. Many patients receive concomitant antiarrhythmic drug therapy which is given to prevent frequent episodes of sustained ventricular tachycardia or nonsustained ventricular or supraventricular tachycardia which might cause the device to fire excessively or unnecessarily. In most instances and especially in those patients receiving concomitant amiodarone, attempts are made to keep the dose of the drug as small as possible in order to maintain arrhythmia control with the fewest drug side effects.

COMPLICATIONS

Early in the defibrillator experience silver tinsil lead material was used. This had a high failure rate and these leads have now all been replaced by DBS (drawn braised strands) leads which have had an extremely low failure rate. Since the spring lead exerts considerable tension at the suture site in the the left infraclavicular area, careful attention to the anchoring of this lead is important to avoid spring migration. The use of "butterflies" for suturing is mandatory and has nearly eliminated the problem of SVC spring migration. Several patients have had their arrhythmia accelerated from a tolerated ventricular tachycardia to ventricular fibrillation with inability of the device to terminate the ventricular fibrillation. In our series this represents 2.8% of patients (10). In such patients we have elected to either not implant or to explant the device. Infection is a major complication which has occurred in 3% of our patients and can result in the need for total system explantation including thoracotomy to remove the chest hardware.

FUTURE

Although the currently available AICD device has functioned extraordinarily well and clearly resulted in a striking reduction in the expected sudden death mortality, nonetheless, there will be a number of improvements in the future. A reduction in device size and manufacturability will result from the use of integrated circuits rather than discrete electronic components. Since pace termination or low energy cardioversion will result in energy savings and less discomfort to the patient, these features will be added to the device. Backup VVI pacing will also be added so that the device can deal with bradyarrhythmic events occurring either primarily or as a post shock rhythm disturbance. The use of a transvenous catheter to replace the extrapericardial patch lead will eliminate the need for thoracotomy in many patients requiring defibrillator implantation (15). The incorporation of an ECG memory will permit telemetry of the ECG signal prior to and following each defibrillator discharge and will result in improved ability to determine the cause for clinical shocks received by the patient. Full external programmability of shock energy, rate cutoff for shocks, sensing criteria, and antitachy pacing modes will result in a more flexible and clinically effective device. Finally, these improvements should permit application of the device to other patient populations potentially ultimately including patients at high risk of sudden death, but who have not yet experienced their first clinical event.

REFERENCES

1. Swerdlow CD, Winkle RA, Mason JW: Determinants of survival in patients with ventricular tachyarryhthmias. E Engl J Med 308: 1436-1442, 1983.

2. Davy JM, Soderholm-Difatte V, Freedman RA, Dorian P, Trusso J, Echt DS, Winkle RA: Survival of 168 ventricular arrhythmia patients on amiodarone therapy. (Abstract) Circulation 70 (Supp II): II-436, 1984.

3. Mirowski M, Mower MM, Staewen WS, Tabatznik B, Mendeloff AI: Standby automatic defibrillator. Arch Intern Med 126: 158-161, 1970.

4. Mirowski M, Mower MM, Gott VL, Brawley RK: Feasibility and effectiveness of low-energy catheter defibrillation in man. Circulation 47: 79-85, 1973.

5. Langer A, Heilman MS, Mower MM, Mirowski M: Considerations in the development of the automatic implantable defibrillator. Med Instrum 10: 163-167, 1976.

6. Mirowski M, Reid PR, Watkins L, Weisfeldt ML, Mower MM: Clinical treatment of life-threatening ventricular tachyarrhythmias with the automatic implantable defibrillator. Prog Cardiol 102: 265-270, 1981.

7. Mirowski M, Reid PR, Winkle RA, Mower MM, Watkins L, Stinson EB, Griffith LSC, Kallman CH, Weisfeldt ML: Mortality in patients with implanted automatic defibrillators. Ann Intern Med 98: 585-588, 1983.

8. Mirowski M, Mower MM, Langer A, Heilman MS, Schreibman J: A chronically implanted system for automatic defibrillation in active conscious dogs. Circulation 58: 90-94, 1978.

9. Reid PR, Mirowski M, Mower MM, Platia EV, Griffith LSC, Watkins L, Jr., Bach SM, Jr., Imran M, T omas A: Clinical evaluation of the internal automatic cardioverter-defibrillator in survivors of sudden cardiac death. Am J Cardiol 51: 1608-1613, 1983.

10. Echt DS, Armstrong K, Schmidt P, Oyer PE, Stinson ED, Winkle RA: Clinical experience, complicationds, and survival in 70 patients with the automatic implantable cardioverter-defibrillator. Circulation (In press).

11. Winkle RA, Stinson EB, Echt DS, Caltagcrone P: Practical aspects of automatic cardioverter/defibrillator implantation. Am Heart J 108: 1335-1346, 1984.

12. Winkle RA, Bach SM, Jr., Echt DS, Swerdlow CD, Imran M, Maso JW, Oyer PE, Stinson EB: The automatic implantable defibrillator: Local ventricular bipolar sensing to detect ventricula tachycardia and fibrillation. Am J Cardiol 52: 265-270, 1983.

13. Winkle RA, Stinson EB, Bach SM, Jr., Echt DS, Oyer P, Armstron K: Measurement of cardioversion/defibrillation thresholds i man by a truncated exponential waveform and an apical patch superior vena caval spring electrode configuration. Circulation 69: 766-771, 1984.

14. Mead RH, Echt DS, Stinson EB, Schmidt P, Winkle RA: The automatic implantable defibrillator: Improved defibrillation an lowered impedance using two large patch leads. (Abstract) Am Coll Cardiol (In press).

15. Winkle RA, Mead, RH, Echt DS, Stinson EB, Bach, SM, Jr.: Automatic implantable defibrillator: Comparison of catheter defibrillation and defibrillation using superior vena spring patch electrodes. (Abstract) J Am Coll Cardiol (In press).

19

NEW ANTIARRHYTHMIA DEVICES - FDA'S REQUIREMENTS FOR EFFECTIVENESS

Glenn A. Rahmoeller

Summary

Many antiarrhythmia devices are now either being marketed or are under clinical investigation in the United States. These devices include traditional pacing systems such as inhibited single and dual chamber pacemakers (AAI, VVI, DVI and DDD) for the control of bradyarrhythmias and for overdrive suppression of certain tachyarrhythmias. The newer burst pacing systems and implantable cardioverters are being studied for control of both supraventricular and ventricular tachycardia and for control of ventricular tachycardia and fibrillation. To establish effectiveness, FDA requires that the manufacturer demonstrate that the pacing system provides the benefits claimed in the labeling. The data required to demonstrate the effectiveness for each of these types of pacing systems are discussed.

Background

Pacemaker pulse generators and leads have been used to benefit patients with cardiac conduction defects for about 25 years. The traditional pacing systems (pacemakers) include atrial inhibited (AAI) and ventricular inhibited (VVI) pacemakers; and while new pacing systems such as the DVI and DDD pacemakers provide the advantage of more nearly physiologic pacing, their primary purpose is basically the same as that for the single chamber systems--electrical control of bradyarrhythmias. These traditional pacemakers are, however, also used for overdrive suppression of certain tachyarrhythmias.

It is the newer burst pacing systems and implantable cardioverters and defibrillators that are being investigated primarily for the electrical control of atrial and ventricular tachyarrhythmias and ventricular fibrillation. These devices offer potentially significant benefits--some truly life-saving--to many patients. However, extensive electrophysiologic (EP) testing of the patient and evaluation of the pacemaker during EP testing is generally required before these antiarrhythmia devices can be evaluated in long-term clinical studies.

What is Effectiveness?

Before the requirements for effectiveness for these various pacemakers can be described, the term "effectiveness" should be defined. For the purpose of this presentation, effectiveness means the ability of the pacemaker to sense and pace as designed, and to provide the benefits that are described in the indications in the labeling. This also means that the labeling must define which patients will benefit from the device, and specifically what those benefits are.

Traditional Pacing Systems

Basic indications for these types of pacing systems for bradyarrhythmia control have existed for many years; according to the Joint American College of Cardiology/American Heart Association Task Force on Assessment of Cardiovascular Procedures-Subcommittee on Pacemaker Implantation (ACC/AHA Task Force)[1], there is general agreement that pacemakers should be implanted for this purpose. FDA's requirements for the showing of effectiveness of these pacemakers for this purpose has generally been based on clinical evaluations by physicians with experience with these devices.

FDA's requirements for clinical data for new dual chamber pacing systems generally includes 3 to 4 month follow-up (depending on the intensity of the evaluation) on 100 pacemakers. The database generated by the protocol suggested by the North American Society of Pacing and Electrophysiology (NASPE)[2] includes an intensive 30 day evaluation of 30 pacemakers implanted by 3 to 5 pacemaker experts, an additional 60 days of follow-up of the first 30 implants, and a less intensive 30 day evaluation of an additional 70 implants. The NASPE protocol also includes a detailed description of the follow-up data and the training and experience expected of clinical investigators. This database has been used by several manufacturers in applications approved by FDA.

In addition to the use of traditional pacemakers for bradyarrhythmia control, these devices have also been used for overdrive suppression of certain tachyarrhythmias. The ACC/AHA Task Force considers this to be a condition for which pacemakers may be used, but for which there is a diversity of opinion regarding the necessity of a pacemaker implant. It is my impression that this general indication has come about through medical practice by certain pacemaker implanters and, as such, FDA generally relies on the same type of clinical data for this use of a pacing system as described above. That is, FDA has approved multiple indications (i.e., bradyarrhythmia control and overdrive suppression of certain tachyarrhythmias) of new DDD pacing systems based on a single NASPE type clinical trial including data for both indications.

New Antiarrhythmia Devices

The new antiarrhythmia devices include burst pacing systems for the control of tachyarrhythmias and the implantable cardioverter and implantable defibrillator for cardioversion of ventricular tachycardia and fibrillation. These devices are still under clinical investigation to establish their safety and effectiveness. Generally, the studies involving these devices include extensive EP evaluation of the patients to determine that they meet the criteria for entry into the study, and evaluation of the device in the patient during EP testing to show that the device is safe and effective under controlled conditions.

To establish that this type of device is effective, the sponsor must show (a) that the device accurately senses and responds to the arrhythmia, or arrhythmias, that it is intended to control; (b) that it safely cardioverts the arrhythmia; and (c) that it provides the benefits to the patients that the sponsor claims in the indications in the labeling.

With regard to accurate sensing, data from EP testing and long-term clinical testing must establish the true positive, true negative, false positive, and false negative sensing rates of the device from the patient population in whom it is intended for use, and under the conditions in which it is intended for use (e.g., with or without concomitant drug therapy). Sensing effectiveness can be recorded during EP testing, but is more difficult to document during long-term clinical testing unless, for example, extensive Holter monitoring is performed.

Similarly, the sponsor of the device must provide data to show that it is effective in terminating the arrhythmia in patients under the conditions for which it is intended for use. Again, this may be relatively easy to show during EP testing, but may be more difficult to document during long-term clinical testing without extensive Holter monitoring. One may, however, when dealing with a life-threatening arrhythmia, be able to show effectiveness by a decrease in mortality if the study includes an appropriate control population or if there exists a well documented and appropriate historical control.

And finally, since the sponsor must show the performance of the device in patients under the conditions for which it is intended for use, the sponsor must establish and obtain FDA approval of a well defined protocol under which he will study the device. The protocol must include reasonably detailed patient selection criteria and specify follow-up data collection requirements. It also may be appropriate to consider phasing clinical investigations for such devices, as some manufacturers have done. The first phase, consisting of a limited study, would be designed to learn more about the device so that a clinical protocol for final testing could be developed. Then, only after this first phase is completed, does the manufacturer conduct a more extensive clinical investigation to establish the safety and effectiveness of the device that will be marketed.

Future Goals

As some of these new antiarrhythmia devices are approved for marketing in the future, FDA will have more specific information about the effectiveness data that are required. Detailed guidance will come primarily with experience after FDA, with the advice of its advisory committee (the Circulatory System Devices Panel), has reviewed several applications and has, by practice, established a policy. A policy does exist for the data required to show the effectiveness of the traditional pacing systems; one is yet to be developed for the newer antiarrhythmia devices. In the meantime, it would be valuable to FDA for clinical investigators and manufacturers of the new antitachycardia devices to consider some of the issues described here and to propose practical, innovative methods for efficiently collecting scientifically valid clinical data, as NASPE has done for the traditional pacing systems.

References

1. Joint American College of Cardiology/American Heart Association Task Force on Assessment of Cardiovascular Procedures (Subcommittee on Pacemaker Implantation): Guidelines for permanent cardiac pacemaker implantation. May 1984. JACC 4:434-442, 1984.

2. Griffin, JC: Recommendations of the NASPE policy conference on the clinical investigation of new implantable pacemakers. PACE 6: 159-160, 1983.

PANEL DISCUSSION ON NEW ANTIARRHYTHMIC DEVICES

Dr. Boal: When we are talking about conventional pacing devices, we talk about a device in which it is possible to predict the end of its life. Now we are talking about devices which have high output and which can now prolong a patient's life. Is there some way now that one can look at these, automatic implantable defibrillators and say, we are getting close to the end of the life of it. We need to worry about replacing it. Also are these devices at all proarrhythmic? That is, you are putting a hefty jolt into a piece of myocardium with a pretty fair current density. Are you in effect, after 100 or 200 or 500 shocks going to end up with a big enough scar to generate arrhythmias and my final question for Mr. Rahmoelller, you indicated that these devices should be used only with previous electrophysiologic testing and under careful control. How do you insure that after they have been released to the marketplace?

Dr. Winkle: With regard to the first question, there is a method for predicting end of life, and that is by following the charge time. As the battery function begins to deteriorate the time it takes to charge up to the full 25 joule charge prolongs so that the patients are followed every two months. One will see a prolongation of charge time and that is an indication that the batteries are failing. With regard to the second question about proarrhythmic effects. I don't think anybody knows the answer to that, however, we certainly have done a lot of testing in the operating room and the patient's arrhythmias with several years of evaluation. There does seem to be a flair up of arrhythmias and their not any worse then they were beforehand. There have been some tissue studies done in animals and a few patients who have died with these devices and again there does not appear to be any tissue damage with the current lead configurations. I think it is an important question. If somebody has tissue damage after 300 to 500 shocks, then they probably needed the device and tissue damage is a price that they would have to pay, but everybody has been concerned about that and to the best of everybody's ability to determine the answer, the answer is there doesn't appear to be, certainly no proarrhythmic effect and apparently no long term myocardial damage either.

Dr. Rahmmoeller: It really is up to the manufacturer and investigators as to how they want to propose the marketing of the device and I think that these devices can come to the market earlier, more quickly for clinical use if the restrictions initially are fairly stringent. I think this is what happened with PTCA. If the indications and relative and absolute contraindications hadn't been developed by the PTCA registry, I don't think that therapy would be on the market today. With regard to your specific question about how do we control it after release I think you have to look at what the F.D.A.'s responsibility really is. My view is that our obligation is to assure that there is good science to approve the use of a device under certain restrictions. If I have done that then I think I have done my job. If the physician uses that device or misuses that device, if it becomes generally misused, so that there are a lot of complications, then you will probably find the FDA coming back after the device has been approved and asking for contraindications to be placed on the labeling or more restrictions to be placed on the device. My view is we are in this together and we would like to see the devices get out there and used to help save people as you would and it is your obligation both as clinical investigators and as clinicians after they are approved to work with us as closely as possible on that.

Also I would like to point out that with the traditional types of pacing systems, the single and dual chamber systems, that indications have developed and they are in the literature. They are well recognized in time. The FDA sort of jumped into that in the middle, the devices were on the market without FDA approval because FDA approval wasn't necessary early in the game for those single chamber pacing systems. AAI pacing systems have been used for not only arrhythmia control but also for overdrive pacing. We got in the middle of that, too. That is a different situation than when you come up with a totally new technology like PTCA or some of the antiarrhythmic devices that are now coming to the market, where they haven't been out on the market and there is no clearly developed indications and we are looking more carefully and spending more time looking at those indications for those types of devices than we have on some that have existed for some period of time.

Dr. Spielman: Are you saying that if you were a manufacturer trying to make a device which could deliver concertina bursts or some sophisticated therapy that had no real established place in treatment, that you would suggest that people build themselves a little sample box, rather than make a pacemaker that could do it and then test the box in clinical trial and then if that therapy proved to be effective, then actually make it into a pacemaker? Would that be a logical thing to follow or is it O.K. to just free-associate devices, put them all into a nice little metal can and then expect clinical trials to prove that they do or do not work after they are implanted?

Mr. Rahmoeller: If I understand you correctly, I have suggested exactly what you have said to manufacturer's in the past. I think that what has happened, is that manufacturer's, when you develop a pacing system for example, are not dealing with off-the-shelf components so a great deal of money goes into the development of the pacing system. If you put all your eggs in one basket, then it creates a difficult situation for the manufacturer and the FDA because the manufacturer doesn't want to hear that the device isn't approved. I think there should be a two-phase system. I think it is perfectly legitimate and probably more legitimate to use devices as we have said in an initial clinical phase just to see what the best approach is and then after you get that information, then go to the more committed types of devices where you are buing the integrated circuits for use in a final product.

Dr. Sami: With the current implantable defibrillator, it takes about 10 to 20 seconds to charge the defibrillator. Dr. Winkle, have you studied the effect of that one time such as on an an elderly patient with cerebral vascular disease, and what it does to brain ischemia and possible brain damage? The second question is, have you studied in any systematic way the threshold of various antiarrhythmic drugs or how they affect the threshold so that if we have patients who have those devices can we find a guide as to what drug not to use?

Dr. Winkle: We haven't specifically looked at brain function but nobody seems to have dramatically deteriorated after receiving a shock. We have done a very large series of psychological tests in patients both before and after implants and compared the patients mental status from the stand point of depression and other things and have found them to be exactly comparable to a control series of patients undergoing cardiac bypass surgery. So from that standpoint it is well tolerated. Certainly you can induce ventricular fibrillation, say in the EP lab, and patients are unconscious and seizing by the time 15 seconds go by. It is remarkable how quickly people begin to lose consciousness. The vast majority of the patients with AIDS are in a very rapid hypotensive v-tach and do not lose consciousness. In effect, we train all of the patients in doing cough CPR so that if they begin to feel dizzy and light-headed and just get a few seconds warning that they are going out, they maintain consciousness by coughing. On the second question relating to drugs the answer is yes. We have done a lot of animal work. We haven't really started the clinical studies yet, and this has been done by other people at other centers, we have looked at in the dog, using the spring patch lead configuration, we have looked at quinidine and bretylium, neither of which effect defibrillation threshold. We have looked at lidocaine which causes quite a dramatic increase in defibrillation threshold and we have looked at encainide and some of its metabolites which also cause quite a dramatic elevation in defibrillation threshold. The one drug I know is reported by Dr. Fogoros concerns amiodarone.

Dr. Prystowsky: We haven't had experience with the implantable defibrillator, but regarding some of the points concerning drugs and their effects in sustained ventricular tachycardia, I can tell you that we have reviewed our experience with morphologies and we assume that rate would be a factor. It turns out when you look at somewhere about 10 or 15 factors about sustained VT that the only thing we found to date that has affected the ability of the cardioverter to break VT and you have to remember now we shouldn't be so separate in these units because both of us are coming towards a mid-point and both units I am sure are going to have both capabilities. When you are putting one of these units in for VT, that person may have a particular morphology when you test it. That person may have other morphologies of VT that you don't know about and we have had 6 people who have had two distinct monomorphic VT's and four of those patients have had markedly different responses to cardioversion in the temporary environment in the laboratory, whereas one VT was easily broken and the other VT was very difficult to break. Two of them have been concordant responses. So the point is that you may think that you have got the answer in the patient let's say on the VT line. Now if you are putting in 25 joules, it is not going to matter much, but if you are putting in small amounts of current, these are things you are going to be considering in the future. Make sure that you really have tested all the patient's different forms of VT or look for them at least, because the morphology change can indeed effect the threshold of cardioversion.

Dr. Winkle: That is a very important point because in fact the drug that intrinisically may elevate defibrillation threshold for pure fibrillation, may convert a rhythm from a disorganized one that is hard to convert with low energies to one that is very organized and one joule may terminate it. Even knowing what a drug does intrinsically to defibrillation thresholds doesn't really help you predict what will happen in that patient. I can say, that based on really a large experience in a lot of patients, on a lot of different antiarrhythmic drugs, there has not been any consistent trend for patients to be harder to defibrillate when they are on drugs, with the possible exception of amiodarone. But there is also probably the likelihood that the defibrillation threshold's fall a little bit after the time of implant as the leads adhere in, an so forth, so that may be a fact that improves the defibrillation threshold over time.

Dr. Horowitz: Is the incubation period for an antiarrhythmic device going to be similar to or significantly different either shorter or longer than the incubation period for antiarrhythmic drugs, do you think?

Mr. Rahmoeller: Are you talking about the lag time?

Dr. Horowitz: The incubation period. I am not about the bureaucratic processing time. I am talking about the time from conception of the device through testing, and eventual approval. I am not talking asking about how long it will take you to do the paper work. That is not a fair question.

Mr. Rahmoeller: That is a fair question. The one device that we have had experience with was a burst pacemaker that was reviewed, again we don't set guidelines, we see a lot of different applications coming in, we just don't have the staff to establish guidelines on what is and what isn't acceptable. What we generally do, is when we get enough experience in reviewing applications for a device, we put that data out, so a lot just depends on what the manufacturer wants to submit to the FDA and then whether or not we on the advisory committee consider it acceptable. In that case there were a relatively small number of implants, 40 some odd implants, 30 to 35 which have been followed and the advisory committee given the restrictions that were being placed on the device, had recommended approval. The only thing that has held up the approval of that device, was there were some electronic problems that hadn't been reported at the time but we found out about later and it was a matter of getting those data together later and assuring that the reliability of the device was acceptable before we approved it. So it is quite possible that based on a small number of devices followed for a fairly long period of time, some up to 2 years or longer, that that device would have been approved.

Dr. Horowitz: So that assuming that the development of the electronics doesn't take years, the testing necessary for the device would be shorter than that for most pharmaceutical drugs?

Mr. Rahmoeller: I am not certain what it is for drugs, but I think that if the testing is done well, that it can be based on testing up to a year or so depending on what the data are, but the things that have existed with some of the current devices like the defibrillator, is that it has been changed by the manufacturer a number of times, two or three times. The people who are investigating it are changing the types of patients in whom it is being implanted. I am faced with the dilemma now with patient's families and physicians calling me saying my patient needs this device and yet we have to ask for the application to get it marketed, because the research which is going off in one direction is not being separated from the marketing in the other direction. What I would like to see is a marketing application which we are getting and see a decision made on that. I think the same thing is true of the other devices.

Dr. Boal: I was wondering if you might comment on the effect of drugs on sensing, particularly in ventricular tachycardia. Sensing is based on a combination of slew rate and amplitude and the administration of drugs can certainly change slew rates. Do drugs effect sensing and how do you effect sensing in these devices?

Dr. Prystowsky: I could talk about the experience with the cardioverter. We try our best to put the cardioverter in and test it in a patient who is on a stable drug combination they are going to need. I think the underlying theme here is as everyone here realizes is that this is not in lieu of drugs, because then you will be using it every day. You still have to keep them out of their arrhythmia. So we try as best we can to test them in that state. When we have had to have a drug change, so far maybe we have just been lucky, we have still had adequate sensing with the device, but I guess you could theorize a situation where that might change. At the present time we haven't seen it.

Dr. Winkle: I can comment for the automatic defibrillator, that the sensing circuits have an automatic gain control adjustment that automatically adjusts if there is a changing amplitude in the signal and so forth, and because of that it can be sensitive down to signals. I mentioned a few tenths of a millivolt or lower and we have never seen any problem with sensing effect in literally hundreds, 4-,5-,600 episodes with this device, I have never seen a single episode of VT that was not sensed, that was above the rate cut-off of the device. Now that is a problem however, that if you put a patient on drugs, that the drug may slow the VT down below the rate cut-off of the device. Fortunately, that usually means that the VT is hemodynamically well tolerated but not always. That is a bit of a problem at the present time since that aspect of the device is not programmable. But other than that problem, I have never seen drugs affect sensing in any way.

Dr. Horowitz: Dr. Fisher, have you ever seen any problems with burst pacemakers and drugs as far as sensors?

Dr. Fisher: The problems with burst pacemakers in sensing, has been largely fixable by the multi-programmability. For example, the implantable burst pacemakers are the type that Mr. Rahmoeller was just talking about. On the other hand, as drugs are added and withdrawn, sensing can change, but again, I agree with Dr. Winkle that our experience with implanted anti-tachycardia burst pacemakers, the difference seems to be independent of whether the patient is on drugs, that is to say that sensing may be good during sinus rhythm, not good during tachycardia. You usually have to reprogram the unit and it does not seem to be so badly drug sensitive that the units become worthless, it usually means just pushing the button and reprogramming it to a more sensitive setting.

Dr. Larry Griffith: One comment and then one question. In the Baltimore experience, we have put in an implantable defibrillator at the time of endocardial resection in virtually every patient that has undergone this operation. This is because several speakers commented that these were not surgical candidates and our experiences were in November, but I believe the numbers are in 27 patients that have undergone endocardial resection with pre-operative and inter-operative mapping. We have had 5 patients that have the device fire subsequent in a 2-year follow-up out of hospital. Not necessarily can we conclude whether they would have or not, but we in fact are doing combined surgery and implantable defibrillator in Baltimore. The question for Dr. Prystowsky is in terms of the experience in Indianapolis for the internal cardioverter. I would be interested in how many patients you put the device in, roughly how many were screened for the device, the success rate of out-of-hospital termination and any difficulties that you have had.

Dr. Prystowsky: How many have been screened for the device. I can't tell you the exact numbers because it has been opened up to many centers around th country and we'll get a call from an investigator, and say I've got a patient we would like to send you for a permanent. I don't know how many they have personally screened to find that one person, so I can't give you the country experience. We have implanted 11 units as permanent implants. We have teste at our own shop approximately 28 patients and we may be more stringent in our criteria than some of the other places, so I can only tell you that most of the ones that we put in have come through our own group of patients, although I think about a third now have come from elsewhere. In the automatic mode we have about 3 or 4. As I said, this is the only patient who has died, the one who was in the automatic mode and this same individual had multiple cardioversions successfully prior to this one episode. We have a few others in the automatic mode, they have occasionally used the unit appropriately. W have had no patient other than this one who has seemed to used the unit inappropriately.

Dr. Griffith: I think you did answer. Have there been any instances of acceleration of the rhythm?

Dr. Prystowsky: Yes, we won't put it in in a patient like that and we have so few that we put in the automatic mode that I can't tell you if anyone has it. That one or two other patients have had an acceleration it has been recognized and then cardioverted, that is something that I just don't personally know. We test all these people back in the laboratory and one interesting thing has happened. One patient who was doing fine in a manual mode, developed a change in his tachycardia over a 1½ year period to the point now that he is accelerated by the device and it was not a drug change, everything else has been stable. He is not in our opinion anymore a candidate for the manual model of termination and I think these are interesting observations. I am not sure why he has changed, he hasn't had any recent obvious second infarct or heart failure or any of the things you look for, but his arrhythmia, consistently, this has happened over a 6 month period, 2 checks, twice each time, has become different. That would be a concern of ours, obviously in the automatic mode and I think that is why I wanted to stress the precautions that we have currently without having de-fib backup.

Dr. Campbell: Can I return to this aspect of approval for new devices and particularly aspects of approving both the device and getting a handle on the indications. We have discussed these devices as though they are hard-wired options just now. That they are specifically made for a given purpose. It seems very likely however, that with the software programmable devices that all of these options and some that we have not even imagined yet, could be developed and produced by a unit that is a multipurpose unit. How do you view the approval process going for that. In other words, the device itself may be proved for a few indications that exist at the present time, but it can be adapted in the future.

Mr. Rahmoeller: We have had some experience with that with the new dual chamber pacemakers. I guess I am personally concerned about two aspects of it, I am concerned from the standpoint of making sure that the physician can operate it. Some physicians have the capability of programming investigational parameters in the device at the same time that the normal purchaser should not be able to program those special parameters. I am worried to make sure that the physician really can't program the device if he is not an approved investigator. The other aspect is I am concerned about the promotion. I am concerned about manufacturer's putting on their menu in their programmer things that they will use to sell their device before those options are available. We have allowed manufacturer's to have that dual system, but we make them blank out the programmer; those options that are not available generally and only available during clinical investigations so it is not used as a promotional point. Also, I am concerned about the fact of the ethics of it. I think one would run into a problem if a physician knew that his programmer that he bought on the open market, could do some of these investigational things, and he had a patient who he thought that needed to be done to, I can see the investigation just going down the tube, because everybody becomes an investigational subject then. It has been done under fairly tight control.

Dr. Spielman: The thing that is curious to me about some of the software programmable burst pacing devices is that they have built into them therapies that really haven't been tested very well. Let's say you look at some unusual form of delivering bursts which may have been written about in one article somewhere and is now built into this device. What worries me is that there is not enough data to support the use of that mode and then you've got this thing sitting there. It really makes a big problem, because suppose you could software program anything, then the question becomes, since you don't know what a lot of these things do, and what their role or place is, you have to ask yourself, what are they doing in the device anyway and from a manufacturer's point of view, it seems to me that the trend in pacing has always been to give more, more, more. Now to put everything in there so that Medtronic can say I've got 3 things that Cordis doesn't have in its pacing device and visa versa. I think that is very dangerous when you talk about anti-tachycardia modes, because a lot of them have no proven efficacy. It seems to me the smart thing to do is find out whether they work first in the investigational sense, and then if they don't work, reject them. For all I know, burst pacing of a simple asynchronous type is as effective and no more dangerous than triple stimuli that are waltzed through the cardiac cycle continously. There is not enough data to say, and I don't think personally that the devices should be allowed to be released in which the modes that can be programmed have no proven value.

Dr. Prystowsky: I just want to take a dissenting opinion on that. I agree with all the precautions, but I would hate to see things become obsolete before their time. Too many things in this country that we all buy are in that category. There are a lot of very smart engineers in these pacemaker companies who continually call you and say I've got this idea, what do you think about this. It may have no basis at the moment for therapeutic modality but 3 or 4 years from now, as we learn more, we may say that was a terrific idea.

Dr. Horowitz: Should the AID and the internal cardioverter be approved as VVI pacemakers?

Dr. Prystowsky: I won't say anything on that.

Mr. Rahmoeller: I think part of the reason that the manufacturers build the devices the way they do is because if somebody says they want more, more, more, but the physicians have never yelled stop, stop, stop.

The demand appears to be there. The second thing is and I am not advocating this, part of the reason the manufacturers do it is, it is a lot cheaper to put everything in the device to begin with, than it is to go back and change it each time. You are talking about not taking off the shelf components. You are talking about designing the integrated circuits to perform a specific function. But then to flip the coin, there have been devices of where electrocautery, the memory in the device, in the pulse generator was actually changed so that when the pulse generator was interrogated, instead of saying I am a model 3 pulse generator, it told the programmer it was a model 2 pulse generator which may not have even been on the market, so that just brings up an example right now where the device haven't been safeguarded well enough to protect against this, so I am just pointing out dilemmas, not solutions.

Dr. Fisher: One of the major costs of research and development is in fact finalizing the microcircuitry and buying it, and that is a tremendous expense. As I understand it, most of the modern pacemakers presently available from any given company are in fact all the same pacemaker. When you buy a Medtronic, or Cordis or Intermedics VVI in fact is the same pacemaker as their DDD with certain parts of the chips activated and not activated and so forth. The reason for that is that it is cheaper to build a single chip that can be used for a million different things, but with access limited, when it is sold as a VVI than it is to make completely separate components for each pacemaker and the same thing holds with respect to the anti-tachy devices. It is indeed a lot cheaper to make a single multi-capable chip and then in some way limit access to it with the programmer or whatsoever, so that the untried methods are limited. I remember with that pacemaker talked about that has been in limbo for a long time. There are I believe 39 different antitachycardia responses and one of the questions that came up at that panel meeting were how often had each of these 29 options been in fact chosen and whether the unit if approved, should be approved if choice #17 had not been used, or had been used only once, should the unit be approved. Should access be limited and so forth. I believe in that case the decision was to accept the machine with all its choices available.

Mr. Rahmoeller: Dr. Fisher was very persuasive at that meeting as he is here, but I would like to disagree with him and I think I have hard data on it. The devices that you are seeing, the VVI pacemakers, I don't believe are subsets of the DDD pacemakers. I don't know why the manufacturers have chosen to go that way. The reason I am certain about what I am saying is that we are talking to the manufacturer's right now about a requirement in the law that says that at some time all those single chamber pacemakers that have not gone through the pre-market approval process will be required to go through the pre-market approval process. They could very easily supplement their already pre-market approved devices with single chamber units and they haven't done it. I think part of it may because of the regulatory process that they may not want to report adverse experiences that they have to report on under pre-market approved devices. They don't have to report or haven't to date for these other devices that are considered substantially equivalent to Holter model units, but I can tell you there are a large number of single chamber units out there that are not indeed subsets of the dual chamber units that have been approved.

V. EVALUATION OF ATRIAL ARRHYTHMIAS

20

STUDY DESIGNS TO EVALUATE ATRIAL ARRHYTHMIAS ARE EASY

ROBERT TEMPLE, M.D.

A drug can be hard to study for a variety of reasons, including:

1. Patients with relevant disease may be hard to find and/or recruit.
2. The condition may be inherently difficult to assess because of great spontaneous variability, a lack of objective measures of severity, substantial disagreements about what should be measured, or vagueness in definition of patients, such that patients identified for study are very heterogeneous.
3. Good studies, while easy to design, may be difficult to conduct because of ethical constraints, such as reluctance to use placebo control groups or a strong perception, based on open observations, that a drug is effective and that denying it to people is wrong. This is a problem with almost any life-threatening illness that is not well-treated by available drugs.
4. The drug's role is as prophylaxis against relatively rare events, or events that take many years to occur, e.g. studies of various agents in prevention of stroke or myocardial infarction; in fact, almost any mortality endpoint study that does not involve a progressive, rapidly fatal disease, such as cancer or ALS, where a large fraction of the patients die is hard to study.

In several ways, evaluation of the effectiveness of anti-arrhythmic agents in treating atrial and ventricular arrhythmias is comparatively simple, at least up to a point. It is easy because:

1. The events in question are entirely objective and can be measured with great accuracy. With the exception of artifacts and perhaps rare peculiar complexes, it is readily possible to say how many VPBs, APBs, runs of a given size, etc. occurred

in a 24 hour period. So far as we can tell, there is little true placebo effect and investigator bias is eliminated as a possibility because the data come from Holter tapes read by a third party.

2. Patients are relatively stable and drug effects are large. While there is considerable day-to-day variability in most arrhythmias, in many cases they are stable enough that it can be argued that patients can serve as their own control, i.e., that you can compare the treatment period with the patient's baseline. This is a study design we usually reject, as it lacks a concurrent control, but if a drug-related change is large, exceeding the reasonably well-defined variability within a patient, and most antiarrhythmic drugs under study do produce such large changes, a case can be made that the effect can be legitimately observed even without the concurrent control. Even if one balks at this, we regularly accept positive control designs as valid for anti-arrhythmics, so long as the drug effect is reasonably large, beyond what could occur by chance.

   Smaller drug effects, if they are of interest, can be detected by including a concurrent placebo control, or by introducing a drug withdrawal phase, to provide a precise assessment of the patient's spontaneous changes.

   The methods of study are in fact now so good that we can detect changes that, at least on average, are probably not even of clinical interest. This situation is not the usual one. For many drug classes it is reasonably difficult to show any drug effect at all and studies must be rigorously designed with concurrent control groups, usually placebo controls, to do so. You cannot, for example, evaluate antidepressants, analgesics, anti-anginal agents and most other drugs using a baseline control or positive control methodology, because spontaneous resolution, placebo effects, and observer biases are major problems that are not controlled with such designs. Even with good studies, known effective agents in these classes may not be demonstrably better than a placebo in a given trial.

3. The regulatory effectiveness standard for approval has been demonstration of an effect on the arrhythmia, not survival.

Evaluation of drugs for ventricular arrhythmias is made difficult, on the other hand, by two factors:

1. With respect to the common arrhythmias like excess VPBs, whose evaluation is easy, a major purpose of treatment, prevention of a fatal arrhythmia, is not directly assessed by our usual means of measuring the antiarrhythmic response, namely, finding a change in arrhythmia rates. In fact, studies to date have consistently failed to show a reduction in mortality by drugs that did reduce VPB rates. This could imply that many or most of the deaths in patients with arrhythmias are often not the result of an arrhythmia at all, or, more likely, that it is not the result of the arrhythmias we are seeing and treating. Various means have been used to identify a more meaningful set of arrhythmic events and show that these can be suppressed; there has been a great deal of discussion, here and elsewhere, of the role of exercise testing and programmed electrical stimulation in seeking measures of drug effectiveness that might correspond to an improved survival. Nonetheless, whether anti-arrhythmic treatment prolongs life, except perhaps when used to suppress obvious, persistant, and immediately life-endangering arrhythmias, remains an unanswered question. The evidence that antiarrhythmic drugs can provoke arrhythmias as well as suppress them perhaps makes the problem more urgent but does not alter it fundamentally. It remains true that the most common, and most widely treated, ventricular arrhythmia, frequent VPB's, can be treated successfully, in that its frequency can be reduced, but with completely uncertain effects on mortality. So long as the regulatory standard for approval is an effect on the arrhythmia, new agents can be approved despite this uncertainty. But were the standard to change, and recent discussions by our advisory committee suggest some

dissatisfaction with the current criteria, drug evaluation could be much harder. It already is more difficult for drugs that are not intended to affect the primary arrhythmia, such as so-called "anti-fibrillatory" drugs or drugs that are so toxic that only an effect on mortality would justify their use.

2. When a patient has infrequent, but potentially dangerous arrhythmias, it is difficult to carry out a controlled trial, because use of placebos or known ineffective therapies does not seem ethically acceptable. Unfortunately, while the spontaneous variability of VPB rates has been pretty well studied, much less is known about spontaneous variability of rarer, more serious arrhythmic events. Despite this, in these cases open, baseline controlled studies may be all that can be conducted and it is often difficult to sort out the patient's baseline status, effects of prior therapy, and response to the new treatment. When a drug is intended only for treatment of such serious events, e.g., because it is quite toxic, evaluation can be made difficult. The best current example of this is amiodarone, for which conventionally designed, concurrently trials have not been carried out.

Unlike the ventricular arrhythmias, most atrial arrhythmias are treated for themselves alone, not to prevent some more serious outcome that might be related to the arrhythmia. The measurements we make of their frequency thus are measurements of the real clinical endpoint. This happy situation has a caveat, however, that bears close attention. In treating an atrial arrhythmia we do not want to expose the patient to a still worse risk. Yet some drugs commonly used for atrial arrhythmias, such as quinidine, are arrhythmogenic, and certain patients with atrial arrhythmias may be at a relatively high risk for more serious arrhythmias. We thus need to be concerned about a potential adverse effect of effective treatment on survival, perhaps especially in patients with structural heart disease.

Although atrial arrhythmias can be annoying, frightening, and debilitating and can have serious hemodynamic consequences while present, each episode is in most cases not so dangerous that they cannot be studied. This contrasts with, e.g. ventricular tachycardia or fibrillation where, at least in outpatients, the first episode during a trial may well be the last. It is therefore usually ethically reasonable to carry out randomized, controlled trials in virtually all atrial arrhythmias, and, as noted, if we find a favorable effect it usually corresponds to a real clinical benefit.

Let me now consider several specific arrhythmias and the evaluation of drugs to treat them.

1. Acute conversion of supraventricular tachycardia

   If you can find the patients, this is a very easy indication to study. It is especially easy when the drug is used in one or two doses. We have in this case a relatively recent approval as guidance. Knoll's verapamil was approved in 1981 for conversion to sinus rhythm of paroxysmal supraventricular tachycardias, including those associated with accessory bypass tracts.

   Studies were carried out in both EP-induced and spontaneous SVTs, both using the same design. (Figure 1). Patients were not stratified according to the electrophysiologic mechanism of the arrhythmia but the studies included patients with and without accessory bypass tracts. Patients whose induced tachycardia lasted at least 10 minutes or whose spontaneous tachycardia lasted long enough to reach the investigator, and whose ventricular rate was at least 100 bpm, were randomly assigned to blinded treatment with verapamil 0.075 mg/kg (up to 5 mg) or IV saline placebo, and the response was observed. Non-responders after 10 minutes were crossed over and non-responders after 30 minutes received an additional dose of 0.15 mg/kg (up to 10 mg) open-label verapamil.

   Results are shown in Table 1. Both spontaneous and induced SVTs were converted by low dose verapamil 30-45% of the time, compared to a 5-10% placebo response rate. If the high dose results are added (there is no placebo control for that segment) overall verapamil response was 15/19 (79%) for the induced arrhythmias and 22/40 or 55% for spontaneous SVT. Five patients with spontaneous SVT had a marked slowing of rate to below 100 beats per minute but did not convert to NSR.

FIGURE 1

Verapamil SVT Conversion Study Design

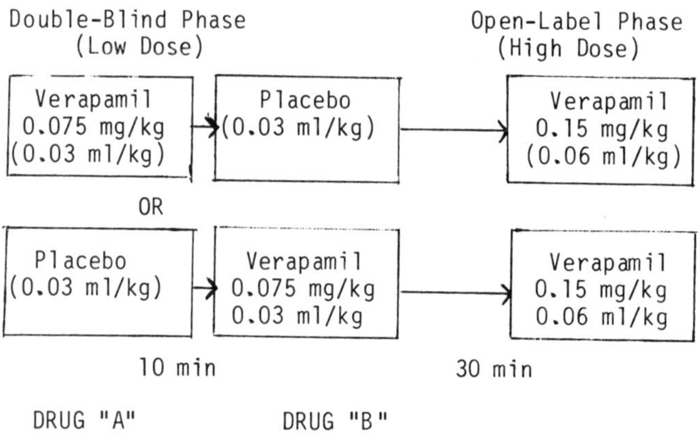

TABLE 1

Verapamil SVT Conversion Study.
Rate of Conversion to Sinus Rhythm

Stimulated Patients - Myerburg/Sung Study

	Double-blind		Total	
	First Injection	Second Injection	Low Dose	High Dose
Verapamil	4/10	5/9	9/19	6/10
Placebo	1/10	0/6	1/16	-

Spontaneous SVT - all studies pooled

	Double-blind		Total	
	First Injection	Second Injection	Low Dose	High Dose
Verapamil	7/20	8/20	15/40	7/16
Placebo	1/21	2/11	3/32	-

Several points should be noted: First, even in patients who have had sustained SVT long enough to reach a medical facility, there was a spontaneous/placebo response in 5-10%, so that a placebo control is advisable. Second, one should not assume spontaneous and induced arrhythmias are equally responsive to an intervention. In this case the induced SVTs seemed more responsive, although admittedly the numbers were small. Third, the crossover design was successful here, with similar results in each period. Fourth, we accepted induced arrhythmias as part of the evidence of effectiveness. Fifth, this study design was accepted, in part, because the response was expected to be rapid and only a short period of placebo observation was needed. Sixth, the dose response aspect of the study would have been better studied if patients had been randomly assigned to placebo, low dose, and high dose. Given the documented, low spontaneous conversion rate, however, the design used was adequate.

2. Acute conversion or slowing of atrial fibrillation

Verapamil was studied in patients with atrial fibrillation at rates of at least 100 bpm using a study design almost identical to the one used to study supraventricular tachycardia. Patients who had at least a 20% fall in rate or converted to NSR were considered responders.

Among the eight studies of reasonable size, involving 134 patients, low dose verapamil produced a mean decrease in heart rate of 25% vs. 8% on placebo. High dose verapamil also produced a 25% reduction (but note only non-responders got the high dose). The 20% fall response rate was 62% on low dose verapamil vs. 15% on placebo. Conversion to NSR was not common: 8/124 (6%) on verapamil and 5/104 (5%) on placebo. The presence of a placebo control was essential to showing that verapamil does not convert atrial fibrillation to NSR, although open studies in Europe had attributed a low conversion rate to it.

Studies of conversion of atrial fibrillation to sinus rhythm, if response was expected to be delayed, probably should use a straightforward parallel, concurrent control design with conversion rate the endpoint.

3. Chronic suppression of paroxysmal arrhythmias

   a. APCs

      Treatment of APC's is a minor matter, as they ordinarily need no treatment. To study drug effectiveness, however, is very simple. Patients with APC's would be randomly assigned to treatment and placebo with Holter monitoring used to assess the reduction in APC rate. It might be argued that symptoms such as palpitations should also be assessed, but as these are so clearly the result of the arrhythmia, reduction in APC rate should suffice.

      While showing effectiveness would be easy, if the drug were a new agent not approved for other uses it would be reasonable to ask whether it could be approved for so minor a use without evaluation of more important uses. Certainly, there would have to be a good deal of evidence of safety.

   b. Controlling rate of chronic atrial fibrillation

      An effect on the rate of chronic atrial fibrillation is readily assessed through a randomized trial comparing the test drug and placebo. Initially, patients with very high rates or patients who tolerated the arrhythmia poorly might be excluded but they might be studied later in trials that allowed early withdrawal for unacceptably high rates, evidence of heart failure or evidence of ischemia. In many cases the initial studies would have to be carried out in patients partly controlled on other agents, such as digoxin. This should not be a problem, and would provide data on an important potential interaction, but it might lead to an initially narrow claim. It is important, and easy, to gain dose response information in these patients. This could be done in a simple, parallel design trial with several groups, each given a single dosage (or titrated up to a pre-determined dosage). A multiple crossover design might also be used and would give dose-response curves in individuals as well as for the group; the crossover sequence should be randomized.

In a recent study of this indication a sponsor used a technique that has become familiar in VPB trials: open titration to an "optimal" dose followed by a randomized crossover trial comparing the chosen optimal dose and placebo.

At least five simple, acceptable designs that could show effectiveness in controlling the rate of atrial fibrillation thus quickly come to mind. Figure 2 shows three designs in which either a single dose level or an optimum level is compared with placebo. Figure 3 shows two designs in which dose response data could be obtained. Any one of these designs could, and perhaps should, examine both resting and exercise rates.

The most obvious measure of drug effect would be heart rate taken at one or more time points, such as at trough and approximate peak blood levels. Alternatively, Holter monitoring would provide 24 hour rate data. Probably any of these approaches are reasonable.

c. Paroxysmal atrial tachycardia or fibrillation

Although patients with these conditions are relatively uncommon, the conditions are persistant and patients with them are known to institutions. Treatment is relatively unsatisfactory and the events are usually not dangerous, so that placebo-controlled trials can be utilized. The difficulties are: 1) that patient's event rates are very variable, 2) that paroxysmal arrhythmias of different etiology might respond differently, and 3) identifying an event may be difficult as most events are resolved without institutionalization.

The latter problem could perhaps be solved by worn devices that record and store a few seconds of an arrhythmia automatically or when the patient turns them on, or alternatively, by transtelephonic techniques. You

FIGURE 2: Rate Control in Atrial Fibrillation (Single Dose Level or Optimum Dose)

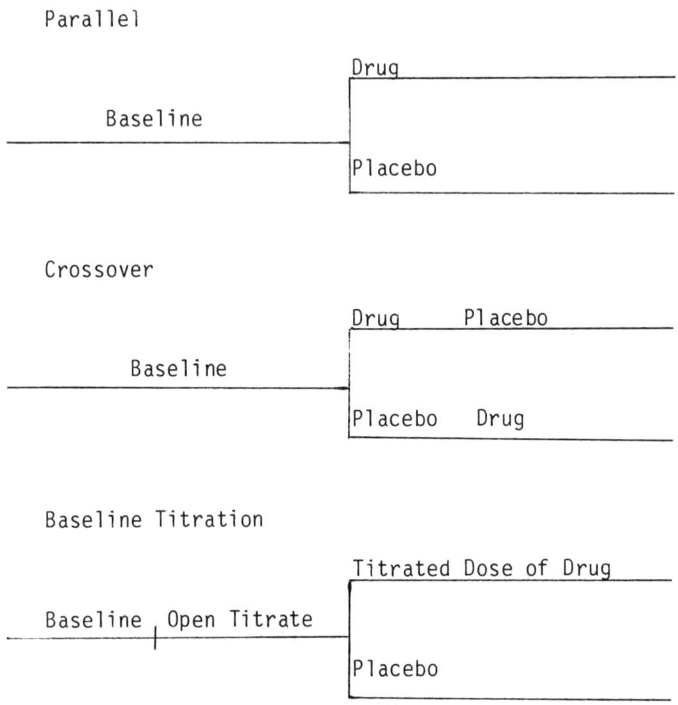

FIGURE 3: Rate Control in Atrial Fibrillation (Dose-Response)

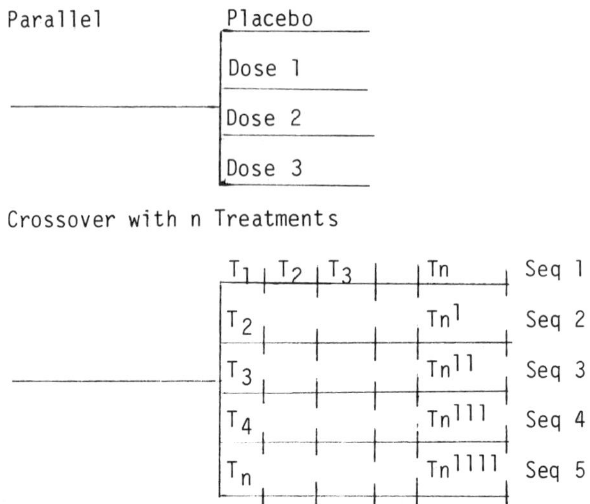

No. of sequences need not equal no. of treatments.

could also rely on diary methods, as these patients are usually able to record their pulse.

Several study designs seem potential useful; in all of them, it is prudent to define the group as narrowly, with respect to etiology, as feasible, in case a particular drug has effects on some, but not other, subsets of patients:

1. If the events are relatively frequent, standard randomized parallel designs or crossover designs seem appropriate, with the endpoints being either frequency of the events on Holter monitoring or, perhaps more important and more relevant, a patient diary recording sustained tachycardia.

   Still better would be a transtelephonic monitoring device or a patient or arrhythmia activated recorder. It could be argued that it is the prolonged events, easily recorded in a diary or transmitted by telephone that are the ones we really want to know about and that a Holter recording of all arrhythmias, even very short ones, might give less pertinent data. Emphasizing the longer episodes would also deal with the possibility that the episodes of tachycardia are fewer but that those that do occur are more sustained.

2. If events are relatively rare in most patients, e.g. less than monthly, and assuming one wants to study chronic therapy in such patients at all, the usual study designs have an unacceptably large chance of failing to find an event in a patient, thus failing to gain information from that exposure. Running the entire study long enough to capture nearly all patients, e.g. 6 months, is one solution, but may be a waste of effort for those patients with relatively frequent events.

   An alternative is to use a design sometimes used in studying anti-seizure drugs, in which a patient's exposure to a treatment is terminated when a seizure occurs and the "time to first event" then becomes the measure of effectiveness. This approach could be fitted into a parallel or crossover design or into dose response designs as well. It would also be a suitable design, indeed the only design, for studying effectiveness of a drug in maintaining patients in NSR after cardioversion of atrial fibrillation.

3. If patients of a particular kind are extremely scarce, a multiple crossover, 2-treatment design could be used, with patients assigned to a random sequence of the two treatments. Analysis of drug effect could be based on the total events or on the time-to-first episode on each treatment in the sequence, a design used successfully to study the rare disease hereditary angioneurotic edema.

It is, as I think has been shown, fairly easy to design trials that can, even with relatively few patients available, assess therapy for atrial arrhythmias, but there is one major concern that needs discussion. Although suppression of the atrial arrhythmia is the main aim of therapy, one does not want to make matters worse. There are reasons to be concerned about several subsets of patients with atrial arrhythmias, and several aspects of a drug's action on atrial arrhythmias merit consideration.

1. Rapid ventricular rates; 1:1 conduction

   A variety of drugs can convert rapid atrial tachycardias with block such as atrial flutter with 2:1 or 3:1 block to a state of 1:1 conduction, with much greater hemodynamic and perhaps ischemic consequences. This has been seen with digitalis preparations, verapamil, and other agents. Labeling for verapamil suggests that its first use in converting an atrial arrhythmia be in a treatment setting with monitoring and resuscitation facilities, including D.C. cardio-version. Enough patients should be studied, in trials of a new agent, to have a reasonable chance of determining whether this risk is present.

2. Ventricular arrhythmias

   Certain atrial arrhythmias are associated with excessive mortality. Kannel, et al [Epidemiologic features of chronic atrial fibrillation, N. Engl. J. Med. 1982; 306:1018-1022] found that mortality in men and women with chronic atrial fibrillation was twice that of controls, perhaps because it occurred in patients with relatively serious myocardial disease.

It also seems possible that treatment for atrial fibrillation can affect mortality. In the 1970's the drug lidoflazine, a antianginal calcium channel blocker that prolongs repolarization, was studied as a drug to convert atrial fibrillation to sinus rhythm. It was rather successful, in fact, at least the equal of quinidine, but some 10% of patients developed new ventricular ectopic beats, including a 3% frequency of ventricular tachycardia or ventricular fibrillation, and a roughly 1% mortality, not in all cases clearly attributable to the drug. The QT prolonging effects of lidoflazine have been observed in many patient populations but studies of the drug as an anti-anginal agent and as a post-AMI agent in some 250 U.S. patients and well over 1,000 European patients in general did not reveal such arrhythmogenicity, although isolated cases of ventricular arrhythmia, especially of the Torsade de Pointes type, have been seen during marketing of the drug for angina. The relatively benign results in coronary artery disease patients compared with the AF patients is not easily explained, but recent work by Somberg may be pertinent. He showed in dogs made chronic atrial fibrillators with aconitine, that atrial fibrillation increased susceptibility to PES induction of ventricular arrhythmias. He did not find evidence that lidoflazine enhanced this increased suceptibility.

While the situation is somewhat murky, it is clear that drugs studied for atrial arrhythmias must be studied for potential pro-arrhythmic effects as well as for their direct effects on the atrial arrhythmia.

No drug evaluation proves as easy in practice as it does in theory and new knowledge inevitably poses new question. Nonetheless, the evaluation of drugs for the treatment of atrial arrhythmias seems, compared to most new drug evaluation, to be straightforward. The main difficulty is probably case finding, so that studies will likely need to be multi-centered, with the logistic problems that poses. Treatment of atrial arrhythmias has not been an area of major pharmaceutical company interest, verapamil being the only drug studied extensively in recent years, so far as I am aware.

21

ROLE OF THE AUTONOMIC NERVOUS SYSTEM IN THE GENERATION OF SUPRAVENTRICULAR TACHYARRHYTHMIAS: CLUES TO DRUG SELECTION

AUGUST M. WATANABE, M.D.

In this brief review the role of the autonomic nervous system in the generation of supraventricular tachyarrhythmias will be dealt with under three headings: first, the complex interaction between the sympathetic and parasympathetic limbs of the the autonomic nervous system will be reviewed; second, in light of that information, the application of drugs that can modify autonomic nervous system function in the treatment of supraventricular tachyarrhythmias will be considered; and, finally, the anticholinergic actions of some class I antiarrhythmic drugs will be discussed with specific reference to the undesirable effects that these anticholinergic properties might have in treating patients with supraventricular tachyarrhythmias.

AUTONOMIC REGULATION OF CARDIAC FUNCTION

The cardiovascular system is regulated in a dynamic, integrated fashion by the autonomic nervous system (1). In general, activation of the sympathetic nervous system is excitatory whereas activation of the parasympathetic nervous system tends to be inhibitory. When both the sympathetic and parasympathetic limbs of the autonomic nervous system are activated simultaneously, there is a complex interaction between the two limbs in which the parasympathetic system modulates the effects of sympathetic activation (1). Thus, the cardiovascular effects of simultaneous sympathetic and parasympathetic activation are not simply a net result of the actions of the two limbs. Rather, parasympathetic effects can entirely override sympathetic effects so that the net cardiovascular response is clear-cut inhibition. Furthermore, under certain conditions, particularly in ventricular tissue, the effects of activating the parasympathetic nervous system are actually magnified during simultaneous sympathetic activation, a phenomenon which has been termed "accentuated antagonism" by Levy (1). This phenomenon of accentuated antagonism can be shown in intact animal studies as well as in isolated organ or tissue preparations.

The interaction between the sympathetic and parasympathetic nervous systems occurs at two levels (1). First, acetylcholine released from vagal nerve endings can interact with muscarinic receptors on sympathetic nerve terminals to inhibit the release of norepinephrine (2). In addition to this prejunctional level of interaction, muscarinic stimulation can modulate cellular responses to beta-adrenergic stimulation at the cellular level (3). Accordingly, muscarinic agonists can modulate the inotropic, metabolic (e.g., activation of phosphorylase), and electrophysiologic effects of beta receptor stimulation on cardiac tissues (3).

Examples of this postjunctional type of interaction in regulating electrophysiological properties of cardiac tissues have been obtained in in vitro studies conducted on canine cardiac Purkinje fibers. It has been shown that acetylcholine can attenuate isoproterenol induced shortening of action potential duration in paced cardiac Purkinje fiber (4). In another model, utilizing cardiac Purkinje fibers, acetylcholine abolished isoproterenol induced slow responses (4). Slow responses induced by zero sodium high calcium medium was uneffected by muscarinic agonists (4). Similar interactions between the sympathetic and parasympathetic stimulation can be shown in atrial tissue preparations studied in vitro. There is also extensive evidence for interaction between the two limbs of the autonomic nervous sytem in intact animal studies (1,3). Thus, sympathetic influences on the heart can be reduced by withdrawal of sympathetic tone and/or by increases in parasympathetic tone. In the latter situation, inhibition of cardiac function can occur with or without withdrawal of the sympathetic influences on the heart. This complex interaction between the two limbs of the autonomic nervous system must be kept in mind when considering the effects of autonomic activity on supraventricular tachyarrhythmias and also in interpreting the effects of drugs which might influence the tachyarrhythmias.

EFFECTS OF DRUGS ON SUPRAVENTRICULAR TACHYARRHYTHMIAS

In considering the influence of the autonomic nervous system on supraventricular tachyarrhythmias, it is necessary to focus on the specific arrhythmia under consideration, because the influence of autonomic tone may be variable depending on the type of arrhythmia considered, and it is necessary to deal separately with the initiation of the arrhythmia on the one hand and with the issue of perpetuation and termination of the arrhythmia on the other hand.

The influence of autonomic tone on the initiation of paroxysmal supraventricular tachyarrhythmias and atrial flutter is only partially understood, although evidence suggests that enhanced sympathetic tone would tend to predispose toward the development of these two types of supraventricular tachyarrhythmias. By contrast, enhanced parasympathetic tone might lead to the development of atrial fibrillation in certain patients and in certain animal models of atrial fibrillation. There is a group of patients who has paroxysmal atrial fibrillation in which increased parasympathetic tone seems to play an important role in initiation of these arrhythmias (5). This syndrome occurs more commonly in men and frequently the arrhythmia occurs at night or after the patient has eaten a meal (5). Often the paroxysmal atrial fibrillation is preceded by a period of sinus bradycardia, reflecting the increased vagal tone (5). Further evidence that the arrhythmia might be precipitated by enhanced parasympathetic tone is the observation that these arrhythmias can be prevented by anticholinergic agents. It is also well known that in open chest dogs, increased parasympathetic tone predisposes the animal to the development of atrial fibrillation by reducing the fibrillation threshhold. Thus, in initiation of supraventricular tachyarrhythmias, the influence of autonomic activity may be variable depending on the type of arrhythmia under consideration and on the model or the clinical situation.

Autonomic influences in perpetuation of supraventricular tachyarrhythmias will be considered in two categories, first termination of the arrhythmia and secondly control of the ventricular rate in the presence of these arrhythmias. In the case of paroxysmal supraventricular tachyarrhythmias (PSVT), the arrhythmia is commonly a re-entrant arrhythmia which utilizes the atrio-ventricular (AV) node as one of the limbs in the re-entrant loop (6). Because the AV node is richly innervated by vagal fibers, activation of the parasympathetic nervous system, which decreases conduction through the AV node, is often effective in terminating the tachyarrhythmia (6). Elevated parasympathetic activity blocks conduction in this limb of the re-entrant circuit and thereby the arrhythmia is abolished. Thus, activation of the parasympathetic nervous system is very effective in cardioverting paroxysmal supraventricular tachycardias.

There are a number of well known reflex means of increasing vagal tone to cardiovert patients with PSVT. These include the utilization of pressor agents to activate the baroreceptor reflex, application of carotid sinus pressure, utilization of the Valsalva maneuver by the patient, and application of the diving reflex. In all of these situations, the ultimate objective is to increase transiently vagal tone so that conduction through the AV node is blocked and the arrhythmia thereby terminated.

Vagal tone in the presence of supraventricular tachycardia can also be increased pharmacologically. This is most commonly done with cholinesterase inhibitors (e.g., edrophonium) (7) and by the administration of cardiac glycosides which produce as one of their effects augmented vagal tone. Because increased sympathetic tone augments AV conduction, another way of approaching the treatment of PSVT is to inhibit sympathetic tone. This is most commonly done by the administration of beta receptor blockers (6).

When treating patients with atrial flutter or atrial fibrillation, the application of increased parasympathetic tone might not convert the arrhythmia to normal sinus rhythm but it may be effective in slowing the ventricular response by decreasing the number of impulses that cross the AV junction. Thus, increased vagal tone can slow the ventricular response both to atrial flutter and atrial fibrillation. Frequently, increased vagal tone will convert atrial flutter into atrial fibrillation. Thus, in the case of atrial flutter and atrial fibrillation, the objective is often to control ventricular rate and not necessarily to try to cardiovert the atrial tachyarrhythmia. Ventricular rate can also be controlled by inhibiting sympathetic tone and this is most commonly accomplished by beta receptor blockade. To summarize, in established supraventricular tachyarrhythmias, increased sympathetic tone is usually undesirable because it may increase the rate of the tachyarrhythmia and/or increase the ventricular response. With established PSVT, atrial flutter or atrial fibrillation, it is desirable to augment vagal tone which might cardiovert the arrhythmia, in the case of PSVT, or slow the ventricular response in the case of atrial flutter or atrial fibrillation.

Some of the drugs that are useful in treating supraventricular tachyarrhythmias produce these desirable effects at least partially by altering the activity of the autonomic nervous system. Some are used for acute therapy whereas others are used for chronic prophylaxis against the recurrence of arrhythmia. In the case of acute therapy, edrophonium is useful for cardioverting PSVT or for slowing the ventricular rate in patients with atrial flutter or atrial fibrillation (7). Digitalis is also useful in these arrhythmias, at least part of the action being related to its ability to augment vagal tone. Calcium antagonists, particularly verapamil and diltiazem, are also useful acutely by blocking conduction through the AV node in patients with PSVT (8). Calcium antagonists are effective in this setting because of the strong dependence of AV node conduction on slow inward current, but also probably because this slow inward current is regulated by autonomic activity and this is another point at which the influence of increased sympathetic tone can be antagonized.

In chronic therapy for regulating ventricular rate or for prophylaxis againsts arrhythmia recurrence, again many of the approaches involve increasing vagal activity or inhibiting sympathetic activity. Digitalis is useful in chronic preventive therapy, at least partially because it works by augmenting vagal tone. Beta receptor blockers are also useful. Amiodarone, the class III antiarrhythmic agent, prevents recurrence because of its effect of increasing atrial action potential duration (9). However, it is also well known that amiodarone has non-competitive beta blocking effects and this likely contributes to its ability to prevent recurrence of supraventricular tachyarrhythmias and to regulate the ventricular response to these arrhythmias should they recur (10). Calcium antagonists are also useful in chronic prophylaxis against recurrence. Finally, class I antiarrhythmic agents are useful in preventing recurrence presumably because of their effect to prolong atrial action potential duration.

Because of the important role of the parasympathetic nervous system in the initiation and perpetuation of supraventricular tachyarrhythmias, the anticholinergic side effects of certain class I antiarrhythmics should be considered. In the case of vagally induced atrial fibrillation, the anticholinergic effects of drugs such as disopyramide and quinidine might actually be desirable and this "side effect" might participate along with the class I antiarrhythmic activity in preventing the development of atrial fibrillation in certain patients who have this arrhythmia related to vagal activity. In most cases of supraventricular tachyarrhythmias, however, the anticholinergic side effects of drugs such as disopyramide and quinidine are undesirable and must be kept in mind when these agents are utilized in patients who have supraventricular tachycardias. The anticholinergic side effects of these drugs must also be considered when interpreting the electrophysiological activity of these agents on cardiac tissues. Depending on the nature and magnitude of autonomic tone influencing cardiac tissues, the pharmacologic effects of disopyramide and quinidine may be qualitatively different.

In vitro studies performed on isolated canine cardiac Purkinje fibers and spontaneously depolarizing right atrial preparations have demonstrated this phenomenon (11). In isolated canine cardiac Purkinje fibers which are paced at a constant rate, the addition of disopyramide and quinidine prolong action potential duration (11). Thus, the direct class I antiarrhythmic effects of these agents are to increase the duration of the action potential. However, if autonomic tone is applied to the Purkinje fiber to mimic the condition in vivo, disopyramide and quinidine will actually shorten action potential duration because of their anticholinergic activity (11). The reason for this is that disopyramide and quinidine abolish the muscarinic modulation of the beta-adrenergic effects on action potential duration. Similar effects are seen with quinidine. Of course, atropine produces similar effects.

In the case of atrial tissues, the effects of disopyramide and quinidine on spontaneous rate are also qualitatively different depending on the presence of autonomic tone (11). When disopyramide or quinidine are given alone to spontaneously beating right atrial preparations, the beating rate of these preparations slows down (11). By contrast, when these agents are given to right atrial preparations under the influence of muscarinic receptor stimulation, because of the anticholinergic effects of these drugs, the spontaneous rate increases (11). Quinidine and atropine produce similar effects. Thus, in atrial preparations the effects of these class I antiarrhythmic agents is qualitatively modified by the presence of parasympathetic tone influencing the properties of the atrial preparation. These results in in vitro studies are consistent with well known clinical observations with these drugs in patients who have, for example, atrial fibrillation. These agents may actually increase the ventricular response in patients with atrial fibrillation. This is why in patients who have atrial fibrillation it is essential that they be well digitalized or be administered other drugs which will block AV conduction prior to the administration of class I antiarrhythmic drugs which have anticholinergic activity.

The fact that these class I antiarrhythmics produce their anticholinergic effects by interacting directly with muscarinic receptors is documented by biochemical studies on cardiac tissues. The muscarinic agonist-induced elevation of cyclic GMP levels in atrial tissues is abolished by drugs such as disopyramide just as it is by atropine. In addition, the muscarinic receptor can be directly probed with radiolabelled ligands such as ^{3}H-quinuclidinyl benzilate (^{3}H-QNB) and the effects of known anticholinergics such as atropine, and antiarrhythmic agents, on the binding of ^{3}H-QNB with muscarinic receptors examined (12). When these types of experiments are performed, it can be shown that both disopyramide and quinidine are quite potent anticholinergic drugs and that they compete with 3H-QNB for binding to the muscarinic receptor similar to the effects of atropine (12).

In summary, the sympathetic and parasympathetic limbs of the autonomic nervous system interact in a complex manner to regulate cardiac function. Parasympathetic activity can modulate sympathetic effects both at prejunctional and postjunctional levels. Increased vagal tone can inhibit the release of endogenous norepinephrine from sympathetic nerve terminals. In addition, stimulation of muscarinic receptors by acetylcholine can modulate the cellular response to beta-adrenergic receptor stimulation by catecholamines. Knowledge of these interactions can provide insight into the interpretation of the effects of drugs that modify autonomic nervous system activity on the electrophysiologic properties of atrial tissues. The anticholinergic effects of certain class I antiarrhythmic agents, such as disopyramide and quinidine, can importantly modify the electrophysiological effects of these agents on cardiac tissues depending on the nature and the magnitude of autonomic tone influencing those tissues.

ACKNOWLEDGEMENT

I gratefully acknowledge the excellent secretarial assistance of Miss Terri Butcher.

REFERENCES

1. Levy MN, Martin PJ: Neural control of the heart. In: Handbook of Physiology - The Cardiovascular System I. American Physiological Society, Bethesda, Maryland, 1979, pp. 581-620.
2. Muschol E: Peripheral muscarinic control of norepinephrine release in the cardiovascular system. Am. J. Physiol. 239:H713-H720, 1980.
3. Watanabe AM, Lindemann JP: Mechanisms of adrenergic and cholinergic regulation of myocardial contractility. In: N Sperelakis (ed) Physiology and Pathophysiology of the Heart. Martinus Nijhoff, 1984, pp. 377-404.
4. Bailey JC, Watanabe AM, Besch HR Jr, Lathrop DR: Acetylcholine antagonism of the electrophysiological effects of isoproterenol on canine cardiac Purkinje fibers. Circ. Res. 44:378-383, 1979.
5. Coumel PH, Lellerq JF, Attuel P, Lavallee JP, Flammang D: Autonomic influences in the genesis of atrial arrhythmias: Atrial flutter and fibrillation of vagal origin. In: OS Narula (ed) Cardiac Arrhythmias: Electrophysiology, Diagnosis, and Management. Williams and Wilkins, Baltimore, 1979, pp. 243-255.
6. Brugada P, Wellens HJJ: Electrophysiology, mechanisms, diagnosis and treatment of paroxysmal recurrent atrioventricular nodal re-entrant tachycardia. In: B Surawicz, C Pratar Reddy, EN Prystowsky (eds) Tachycardias. Martinus Nijhoff, 1984, 131-157.
7. Watanabe AM: Pharmacological modification of cardiac function: Cholinergic agonists and antagonists. BF Hoffman, MR Rosen (eds) Cardiac Therapy. Martinus Nijhoff, 1984, pp. 95-144.
8. Schwartz JB, Keefe D, Kates RE, Kirsten E, Harrison DC: Acute and chronic pharmacodynamic interaction of verapamil and digoxin in atrial fibrillation. Circulation 65:1163-1170, 1982.
9. Coumel PH, Bouvrain Y: Etude clinique des effets pharmacodynamiques et antiarhthmiques de l'amiodarone. J. Des. Agr. 6:69-81, 1973.
10. Charlier R.: Cardiac actions in the dog of a new antagonist of adrenergic excitation which does not produce competitive blockade of adrenoceptors. Br. J. Pharmacol. 39:668-674, 1970.
11. Mirro MJ, Watanabe AM, Bailey JC: Electrophysiological effects of disopyramide and quinidine on guinea pig atria and canine cardiac Purkinje fibers: Dependence on underlying cholinergic tone. Circ. Res. 46:660-668, 1980.
12. Mirro MJ, Manalan AS, Bailey JC, Watanabe AM: Anticholinergic effects of disopyramide and quinidine on guinea pig myocardium: Mediation by direct muscarinic receptor blockade. Circ. Res. 47:855-865, 1980.

22

APPROACHES TO DRUG SELECTION AND SERIAL DRUG TESTING

John C. Somberg, MD
Vilma Torres, MD

1. INTRODUCTION:

The evaluation of antiarrhythmic therapy is predicated on the availability of effective means of measuring cardiovascular arrhythmias and determining a drug effect. There are a number of techniques which exist for arrhythmia detection and to aid in therapy assessment. One of them is the 24 hour Holter ambulatory electrocardiography. This technique has been quite useful and has been reviewed by a number of investigators. Exercise tolerance testing, drug level measurements, and programmed electrical stimulation are other techniques for arrhythmia therapy assessment. Drug level measurements have been demonstrated by Meyerburg and associates to be useful in determining if a Class I agent such as procainamide has been given to an effective level. Patients who have had prior electrical instability, if treated with an effective level of a class I agent, will show a reduced incidence in mortality. Antiarrhythmic drug level measurements may be very helpful in determining compliance with antiarrhythmic therapy, as well as determining if a population will show a reduced incidence of sudden death. But the technique of drug level measurement does not aid us in identifying who is at risk, or whether the antiarrhythmic agent is really benefiting that particular patient. Thus, other techniques for arrhythmia therapy assessment have been looked at.

Exercise tolerance testing has been one technique that Lown and associates have found useful in eliciting occult arrhythmias. This group has demonstrated in a number of studies the ability to elicit arrhythmias that were otherwise not observed to be frequent using Holter monitoring techniques. A number of groups, including our own, have found that most ventricular ectopy is inversely related to heart rate. As the heart rate increases on the graded exercise test, the ventricular premature beat frequency decreases.

In my experience, about 70% of the patients showed overdrive suppression of ambient ventricular arrhythmias during exercise testing and it is only the minority, usually ischemically mediated arrhythmias, that showed exacerbation during exercise testing. To look further at this question, we undertook studies in 24 patients who had a cardiac arrest and symptomatic ventricular tachycardia who did not have frequent ventricular ectopy on Holter monitoring. All 24 patients underwent a battery of tests we have labeled autonomic profile. The autonomic profile consists of upright tilt, cutaneous exposure to cold, isotonic (treadmill) exercise, as well as isometric (hand grip) exercise. Using all 4 tests, only 3 patients showed provocation of 3 beat VT while the other 21 patients showed no ectopy brought out by the exercise test as well as the other tests in the autonomic profile battery (fig 1). This complete lack of ventricular arrhythmias was in marked contrast with the fact that all 24 patients had inducible monomorphic sustained ventricular tachycardia using the technique of programmed electrical stimulation. Although Atkins and associates have reported that isometric exercise may produce arrhythmias not elicited by isotonic exercise (12), the autonomic profiling technique in our hands was not able to provoke arrhythmias that were elicited by the invasive catheter technique. It is important to emphasize that these were not patients who were suspected of electrical instability, but those proven to have electrical instability (having a cardiac arrest) not associated with myocardial ischemic. This patient population was at very high risk for recurrent life threatening episodes and needing antiarrhythmic therapy. Non-invasive techniques to assess antiarrhythmic therapy was not effective in this population.

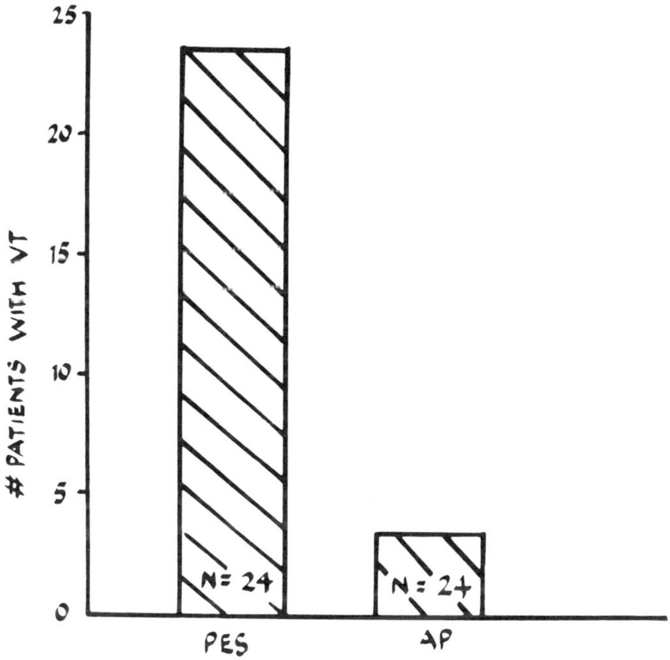

Figure 1. The induction of ventricular tachycardia (VT) by programmed electrical stimulation (PES) is contrasted to the incidence of VT induction following autonomic profile (AP). Positive PES probability of positive AP, 0.208 (95% confidence limits 0.068-0.407). No significant association.

Controversies exist in the use of exercise testing as a provocative intervention for arrhythmia assessment. Are arrhythmias secondary to ischemia being elicited instead of the primary electrical instability?

It may be more appropriate to use antiischemic medications to treat ischemically caused arrhythmias as opposed to antiarrhythmic agents. We must also ask a question, are we selecting the pharmacologic agents that have antiischemic properties instead of antiarrhythmic drugs if we use exercise testing to elicite occult arrhythmias? Furthermore, the reproducibility of these studies is in question, as well as the fact that we are selecting a population who can exercise and who do develop arrhythmias on exercise. This may be an exceptionally small segment of the population, not making this technique useful in identifying the segment of the population at risk for sudden death and thus the most appropiate group in which to study antiarrhythmic drugs.

Invasive Assessment of Ventricular Arrhythmias

The invasive technique of programmed electrical stimulation (PES) may help to determine if electrical instability is present and the ability to sustain ventricular tachycardia. It also helps to define the efficacy of antiarrhythmic therapy, as well as determining the effects of the antiarrhythmic therapy on cardiac electrophysiology. Determining the effects of antiarrhythmic agents on electrophysiology may be very important in that certain agents might depress the cardiac conducting system, requiring concommitant use of a pacemaker. This type of information can readily be obtained from the invasive electrophysiology studies.

Programmed electrical stimulation entails the introduction of pacing trains that are generally 10 to 15% faster then the patients intrinsic heart rate on the initial day of the study. By using a conditioning pacing train, heart rate will not vary from day to day in the study. After the pacing train, usually of 6 or sometimes 8 beats in length, an extra stimulus S2, is introduced. This stimulus is then moved closer to the end of the pacing train at 20 millisecond decrements until it no longer conducts. The point of which it last conducts is termed the refractory period of the right ventricle and the stimulus is either left at the refractory period or placed beyond (10-50 msec) the effective refractory period to insure repeat capture. The stimulus S2, is left at this point, and a second extra stimulus, S3, is introduced as the sum of the S1-S1, S2-S1, intervals. This extra stimulus is then brought in at 20 millisecond decrements until it no longer conducts. In this way, 1, 2, 3, and sometimes 4 extra stimuli can be introduced to elicit the ventricular arrhythmia. Studies performed in patients who have sustained a cardiac arrest have reported a sensitivity of over 90% as well as a high specificity. The studies performed to date reveal that there is no increased risk to the patient beyond the risk of a right heart catheterization.

A number of alternatives exist as to the method of arrhythmia induction. One may reduce the intertrain interval, making it shorter and shorter, therefore increasing the ability to provoke sustained ventricular tachycardia. Also, one can quicken the pacing rate by attenuating the pacing train cycle length. This can decrease the diastolic filling time of the left ventricle, decrease cardiac output leading to ischemia, which would certainly facilitate the induction of VT. Rapid pacing permits earlier introduction of the premature stimulus at the end of the pacing train which may have the properly timed interval to induce sustained ventricular tachycardia. Increasing the duration and the amplitude of the extra stimuli may also greatly enhance the ability to induce ventricular tachycardia. Introduction of the third and fourth extra stimulus may also enhance the induction yield. It is important to realize, though, that not all these alternatives are equally appropriate. I believe that techniques that may lead to ventricular ischemia, such as attenuation of pacing train cycle length, and reduction of intertrain intervals, are less appropiate because they may cause ischemia and ischemia mediated arrhythmias, as opposed to electrical instability not in the presence of ischemia. Furthermore, increasing duration and amplitude of the stimuli, may be a form of fibrillaiton threshold determination and may thus be less useful in determining the prevention of reentrant tachycardia. Timed extra stimuli to elicit a reentrant ventricular tachycardia may be a more sensitive discriminator of electrical instability. Thus, pacing faster and faster at shorter and shorter rest periods will produce results in a serial drug testing program that shows a lack of efficacy for many antiarrhythmic agents requiring techniques such as arrhythmia control pacemakers, implantable defibrillating devices, and/or surgery, as oppossed to pharmacologic therapy. The introduction of a third and fourth extra stimulus, on the other hand, may increase the yield for arrhythmia induction, but not lead to ischemia and the ensuing problems of a patients more difficult to protect with antiarrhythmic drugs.

When evaluating therapy there is a great need for a reference agent if you will an "internal standard". This "internal standard" permits the evaluation of the drug responsiveness of a population. One laboratory may be evaluating patients who are refractory to all pharmacologic therapy while other laboratories are studying patients responsive to both conventional and investigational antiarrhythmic therapy. Knowing that a conventional agent such as procainamide is effective in 50% of patients in one lab as opposed to 5% in another permits the results of the two laboratories to be put in proper perspective. Whenever antiarrhythmic agents are being evaluated the single most important piece of information for me is the response rate in that study to conventional agents. Knowing this the experimental methodology is less important since I can put the very favorable or unfavorable results in perspective. Besides the need for a reference compound, there is also a need to understand the etiology of the patients arrhythmias. This is because programmed electrical stimulation has been shown to be more effective in inducing arrhythmias in certain disease states than in others. For instance, patients with atherosclerotic heart disease show a 95% induction of ventricular tachycardia when they have had a serious life threatening event. Patients with mitral valve prolapse show a 75% rate of induction, patients with a myopathy show a 75% induction rate. However, patients with QT prolongation show only a 20% rate of induction, while those who have VT due to coronary artery spasm, invariably fail to have VT provoked.

Pro-arrhythmic Action of Drugs

The concept of arrhythmogenicity of antiarrhythmic drugs is another aspect that must be evaluated in the drug study and selection process. Recently, Velebit and associates have reported that all conventional and experimental antiarrhythmic agents that they have studied have an incidence of worsening ventricular arrhythmias. They have assessed this by noting that these agents increase ambient ectopy frequency. Another area of possible use for the PES technique is identifying patients in which the antiarrhythmic agents may actually facilitate the induction of ventricular tachycardia. Using PES one may define arrhythmogenicity to mean that the agent lowered myocardial threshold making ventricular tachycardia easier to provoke. It is possible that fewer ventricular ectopic beats may be dangerous in this situation, since even a reduced VPC frequency could still elicit VT when myocardial threshold is reduced. A number of indicies may provide information as to possible arrhythmogenicity of an antiarrhythmic drug.

A drug may facilitate induction while induction before drug administration is not possible. A drug may consistently reduce the number of stimuli required for ventricular tachycardia induction and this reduction may represent a lowering of myocardial threshold due to the arrhythmogenicity of the antiarrhythmic agent. Furthermore, the agent may consistently change the VT induction pattern from nonsustained to sustained ventricular tachycardia. Another area of importance is that an antiarrhythmic agent may make pacing termination of the ventricular tachycardia more difficult or impossible, thus, requiring a higher incidence of defibrillation. In fact, a drug may make ventricular tachycardia induction easier while making defibrillation more difficult. A rise in the defibrillation threshold may be a potentially dangerous action of a drug, separate from other characteristics. The incidence of pro-arrhythmic events and their severity should be documented in the evaluation of antiarrhythmic therapy. Certainly, all antiarrhythmics may be pro-arrhythmic at times but some do this with a far greater frequency and with greater severity. This should be taken into account in the evaluation process.

Substrate vs Trigger Mechanism

Electrical instability may be manifested both by VPC frequency as well as myocardial vulnerability. Some patients who have a normal myocardium may have frequent ventricular ectopy possibly due to increased sympathetic tone. These patients do not seem to be at risk for sudden cardiac death and are not inducible in the electrophysiology laboratory. A triggering mechanism does exist in these patients, but the appropriate substrate of myocardial vulnerability is not present. Another group of patients who have undergone an acute myocardial infarction, have markedly lowered their vulnerability.

These patients may not manifest VPC's; sudden burst of sympathetic tone may be the trigger for ventricular tachycardia and fibrillation that we sometimes see without warning in the coronary care unit. An antiarrhythmic drug may lower VPC frequency, however, the drug may facilitate VT induction by lowering myocardial threshold. Antiarrhythmic drugs in other patients may raise ventricular threshold, preventing ventricular tachycardia induction, while not reducing VPC frequency. It is thus important, both to assess VPC frequency as well as to invasively assess myocardial vulnerability using PES.

Supraventricular Arrhythmias

The evaluation of antiarrhythmic therapy for the treatment of supraventricular arrhythmias (SVT) poses considerable problems. While the problem of SVT is a common one, representing perhaps 60% of clinical arrhythmias the population referred to centers usually involved in the evaluation of antiarrhythmic therapy may be small and not representative of the arrhythmias seen frequently by primary care physicians and thus not the type of arrhythmias the drug will be more frequently used to treat. Although supraventricular arrhythmias may be life threatening for the vast majority, the problem is bothersome but not life threatening. The lack of lethal or potentially lethal consequences limits the evaluation process of drugs for the treatment of SVT. The utilization of invasive techniques must be limited to those patients with incapacitating arrhythmias or who have potentially life threatening arrhythmias. The side effects one is willing to tolerate is also different when it comes to agents in the treatment of SVT. While the toxicity associated with amiodarone is minor when the alternative is death, the possible toxicity is prohibitive when the disorder being treated is palpitations (PAT, APC's, etc). In fact, the spectrum of toxicity seen with the conventional agents procainamide, quinidine, and disopyramide is such that the prescription of these drugs is severely curtailed given the enormity of the problem. This reality is reenforced by the high number of first time in and out of hospital prescriptions not renewed. Besides the wide spectrum of severity one also observes with SVT the difficulty of arrhythmia diagnosis and reproducibility. Often patients complain bitterly about palpitations and disabling arrhythmias but it can take months and thousands of dollars to track down the ellusive arrhythmia. The arrhythmias are very episodic and often the patient may have more than one type of arrhythmia further complicating diagnosis.

<u>Current Approach to Therapy</u>

The field of therapy of SVT has seen considerable change. Although there may be several approaches to a given arrhythmia problem, I have outlined in figure 2 the common approach to supraventricular arrhythmias often taken. It is interesting to note that with the introduction of the calcium channel blocker verapamil and to a lesser extent the beta blockers, the treatment has dramatically changed the use of digitalis and cardioversion for acute problems. However, it is in the area of chronic therapy that the need for effective non toxic agents is so clinically pressing. Furthermore, only in the potentially life threatening arrhythmias usually associated with the pre-excitation syndromes does the assessment of chronic therapy require electrophysiology studies and serial drug testing. These techniques increase the reproducibility of the evaluation process. There is no definitive evidence that agents effective in the pre-excitation syndromes will also be optimum in other supraventricular arrhythmias. Until otherwise established each arrhythmia group must be considered separately and a drug evaluated for each category.

FIGURE 2.

SUPRAVENTRICULAR ARRHYTHMIAS

ARRHYTHMIA	ACUTE THERAPY	CHRONIC THERAPY
SINUS TACHYCARDIA	DETERMINE CAUSE	BETA BLOCKERS
PAT	VERAPAMIL	DIGITALIS BETA BLOCKERS
ATRIAL FLUTTER	VERAPAMIL DIGOXIN	DIGOXIN &/OR TYPE I ANTIARRHYTHMIC
ATRIAL FIBRILLATION	VERAPAMIL	DIGITALIS &/OR CLASS I AGENT
AV JUNCTIONAL TACHYCARDIAS	STOP DIGITALIS TREAT TOXICITY	------------
RECIPROCATING TACHYCARDIAS (WPW SYNDROME, ETC.)	VERAPAMIL BETA BLOCKERS	ELECTROPHYSIOLOGICALLY GUIDED THERAPY

FIGURE 3.

SUPRAVENTRICULAR ARRHYTHMIAS

TECHNIQUES FOR ASSESSING EFFICACY

ARRHYTHMIA	NON-INVASIVE	INVASIVE
PAT	PROPHYLAXIS STUDIES: CROSSOVER VS PARALLEL (TRANSTELEPHONIC SYSTEMS)	ELECTROPHYSIOLOGIC STUDIES (REFERENCE AGENTS PROCAINAMIDE & VERAPAMIL)
ATRIAL FLUTTER	--------	ELECTROPHYSIOLOGIC STUDIES REFERENCE AGENTS LONG-TERM FOLLOW-UP
ATRIAL FIBRILLATION	PROPHYLAXIS STUDIES: CROSSOVER VS PARALLEL (TRANSTELEPHONIC SYSTEM)	ELECTROPHYSIOLOGIC: REFERENCE AGENTS FOLLOW-UP
PRE-EXCITATION SYNDROME ARRHYTHMIAS	--------	ELECTROPHYSIOLOGIC STUDIES WITH FOLLOW-UP

ASSESSMENT TECHNIQUES:

Many techniques exist for the assessment of antiarrhythmic efficacy in the treatment of supraventricular arrhythmias. These techniques can be divided into non-invasive and invasive electrophysiologic assessment. The non-invasive techniques include Holter monitoring, transtelephonic surveillance, exercise tolerance testing, isometric exercise, and diaries. Perhaps the easiest and most inaccurate way to non-invasively assess therapy is the use of diaries and anadotal patient reports. This is helpful and should not be discarded as ajunctive to a more quantitative assessment of arrhythmias, but unfortunately, patient perception and arrhythmia frequency are not directly related. There is also great potential for investigator bias and center to center variability in reporting techniques further weakening this technique for drug assessment. Holter monitoring is of limited value in the assessment of drugs efficacy in surpressing supraventricular arrhythmias because of the infrequency of the arrhythmias and thus the difficulty to capture on a 24 or 48 hour monitoring. The infrequency of the arrhythmias plus the even greater variability of supraventricular arrhythmias then ventricular arrhythmias makes Holter monitoring of limited value. Recently, the use of event monitors which are patient activated has been introduced. However, these systems, although an improvement, still have a limited period of time they can be worn (2 to 5 days) and depend upon the patients awareness of the arrhythmia and ability to respond during the course of the arrhythmia to activate the recording device. It is conceivable that a drug may not decrease the severity or the frequency of the arrhythmia, but somehow incapacitate the patient from responding or decreasing this perception of the arrhythmia, thus making the number of events recorded on the drug less frequent, although in actuality they have become more frequent than during the control period.

Perhaps the most useful recently introduced technique is transtelephonic monitoring. This entails training the patient to use a device that turns the electrocardiogram into an audio signal that can be transmitted over telephone lines and recorded at a central station. These devices can be very useful in permitting the diagnosis of episodic arrhythmias and permit frequent checks on the patients receiving antiarrhythmic therapy. Once again, the techniques are limited by a need for patient perception of the arrhythmias as well as a long enough duration to permit telephonic monitoring. A lot of consideration has been addressed by some of the more recent models that have a memory permitting recording of the arrhythmia and then at a later date permitting the arrhythmia to be played back over the telephone to a central station. Exercise testing may also be of value in provoking supraventricular arrhythmias associated with activity and acceleration in heart rate. The use of an autonomic profiling technique which would include both isotonic (treadmill exercise) and isometric (handgrip exercise), tilt table and cutaneous exposure to cold, may be a useful adjunct to transtelephonic monitoring providing a non-invasive provocative maneuver as well as long-term patient surveillance.

The assessment of therapy must be based on the stratification of the patient population as to etiology of the arrhythmia and the severity of the underlying heart disease. Perhaps most important is the evaluation of atrial size. We know that in atrial fibrillation the perpensity to remain in sinus rhythm is inversly proportional to the left atrial dimension. The frequency and severity of many supraventricular arrhythmias is worsened with increased atrial size. Thus, any evaluation of the drug must take into account atrial dimension and stratify the patients accordingly. Currently, the best technique for atrial dimension evaluation is 2D echocardiography and it is imperative that all patients involved in studies to define antiarrhythmic efficacy in supraventricular arrhythmias undergo echocardiology. The randomization of the control versus intervention population as to atrial size is an appropriate approach to account for this important variable.

The invasive assessment of antiarrhythmic agents offers the advantage of great reproducibility of arrhythmias, repitity of testing and consistancy of results. The use of invasive electrophysiology is limited since only the severest of supraventricular arrhythmias are evaluated invasively. Unfortunately, due to the few patients evaluated in each center, the study population tends to have very heterogeneous arrhythmias. Two patients with re-entrant atrial tachycardia, one patient with pre-excitation syndrome, and two people with proximal AF does not constitute an appropriate study population. Studies must require the hemogenicity of the study population to insure the applicability of the results to a given problem. Patients with pre-excitation syndrome have unique difficulties and the drugs effectiveness in these particular circumstances should be determined, but determined separately from patients with SVT due to other etiologies. Electrophysiologic studies many times produce results that are difficult to be interpreted from center to center due to different methodologies. It is not necessary to legislate the methodology, but rather demand that standard terminology be employed and that an "internal standard" or reference agent in each lab be employed. The reference agent such as procainamide or quinidine could easily make the results of one lab comparable to that of another lab. If procainamide is totally ineffective in one laboratory, while very effective in another, the responsiveness of the population and ease of finding an effective drug given the techniques in a particular center would be appreciated. Once again, it is important to stratify the population studied using 2D echocardiography or left atrial dimension. Far too often, because invasive electrophysiology studies are performed and the results are more reproducible the stratification for atrial dimension is omitted, and thus the long-term results on chronic therapy may be misleading because the propencity to have recurrence is inversely proportionate to left atrail dimension. Methodology for invasive electrophysiology must also be assessed. The use of time extra stimuli may be a more specific and reproducible provocative maneuver than rapid atrial burst and ramp pacing which is far more non-specific. Burst and ramp pacing may induce atrial fibrillation, for instance, in many normal patients (alibie, usually non-sustained). Unfortunately, many experimental protocols mix these techniques while someone may have the tachycardia provoked by timed extra stimuli and following therapy, burst atrial pacing or ramp atrial pacing will be employed. This makes the comparison of results difficult, if not impossible. It also is important to obtain antiarrhythmic levels at the end of the testing procedure and if the testing is prolonged, maybe at the beginning, middle and end of the procedure would be useful. Many times the levels obtained acutely are quite high and not applicable to the chronic administration of the drug. Long-term electrophysiologic evaluation may be needed in patients being treated clinically with drugs that may be metabolized to active metabolights. Some studies in the development program should look at the comparability of test results between the intravenous and chronic oral forms of the drug. If at all possible comparability should be demonstrated, since the acute testing using the intravenous form is far easier and more rapid, and should be favored, if possible.

METHODOLOGIES OF APPROACH

The treatment of proxismal atrial tachycardia (PAT) may be evaluated by both noninvasive and invasive means (Figure 3). Noninvasive assessment usually involves prophylaxis studies where chronic administration of the antiarrhythmic agent being evaluated is compared to a placebo or active therapy (conventional drug). A transtelephonic surveillance system may be employed along with Holter monitoring and provocative testing using at least isometric (treadmill) and possibly isotonic (hand grip) exercise. A parallel study is usually employed due to the episodic nature of the condition being studied and the tremendous variability in arrhythmia frequency. Crossover studies are not feasible with this condition, since there is nothing to insure that a patient weeks or months later will have the same type and frequency of arrhythmias as during the previous portion of the study. Usually this can be solved in crossover studies by making the period of observation shorter, but given the infrequency of the arrhythmias this is not feasible with PAT. Usually the evaluation of drugs for the treatment of PAT does not involve electrophysiologic testing, since few patients with this condition are referred and should be evaluated invasively.

Atrial flutter is very difficult, if not impossible to evaluate noninvasively and too few patients are available for assessment to make a study feasible. A multi-center study approach is probably feasible using invasive provocative techniques with long-term follow-up. Acute studies without long-term therapy and surveillance have little value, since the predictability of drug success on a long-term basis using electrophysiologic testing has not been demonstrated.

Atrial fibrillation is one of the more frequent arrhythmias treated with antiarrhythmic drugs. A drug can be evaluated both invasively and noninvasively. Invasive assessment should involve a referrence agent and have a plan for long-term follow-up. Acute prediction of efficacy must be confirmed with long-term results. Drugs can also be evaluated noninvasively as to the ability to convert atrial fibrillation to sinus rhythm. However, the ability to convert does not necessarily prove that an agent can be effective for long-term prophylaxis. In fact, many agents effective in either controlling the ventricular response in atrial fibrillation or causing conversion to sinus rhythm are not very effective for long-term prophylaxis. Perhaps an optimum approach to the problem is taking the patient in atrial fibrillation who is to undergo cardioversion; administering the drug and then if they don't spontaneously convert with therapy; converting them using cardioversion and following the patients to see how long they remain in sinus rhythm. Stratification as to left atrial dimension, antiarrhythmic blood concentration and type of cardiovascular disease may further enhance the understanding of the efficacy of the agent being studied. These studies are usually performed as a parallel comparison either to placebo or conventional therapy. The use of placebo is very difficult, since most of these patients will require cardioversion and placing someone after cardioversion on placebo may be ethically limited given the high rate of spontaneous conversion back to atrial fibrillation in patients not treated.

Superventricular arrhythmias associated with pre-excitation syndrome should be treated with drugs guided by invasive electrophysiologic assessment. Over the last few years inumerable electrophysiology studies have demonstrated that an agent may have no effect, worsened or improved a potentially life threatening pre-excitation syndrome associated arrhythmia in different patients. For these reasons, experimental assessment of drug efficacy is manditory before long-term therapy is started. Most centers take the approach that patients with electrocardiographic evidence of pre-excitation syndrome without arrhythmias should not be treated. Those who have symptoms or superventricular arrhythmias documented are usually referred for electrophysiologic testing. Certainly to expose these patients to therapy with an experimental agent requires the determination of potential drug benefit that can only be made with acute assessment based on electrophysiology studies.

POTENTIAL FOR PHARMACOLOGIC THERAPY

Currently the field of evaluating drugs for superventricular arrhythmias is becoming more active. A considerable market exists and certainly more drugs are needed to effectively treat superventricular arrhythmias. Efficacy must be determined using many of the newer techniques which permit a more acurate assessment of antiarrhythmic efficacy (Figure 4). A large number of agents under development for ventricular arrhythmias of the Von Williams classification Ia and Ic may also be effective in the treatment of superventricular arrhythmias. Certainly, there still is a need for agents of reduced toxicity that would permit chronic administration to patients who have bothersome, but non-life threatening arrhythmias. For arrhythmias that are potentially life threatening or life threatening many of the Ia and Ic agents under development may be useful and should be vigorously tested with both invasive and noninvasive techniques. Class II agents, beta blockers have had a tremendous commercial development in the last two decades. We still, unfortunately, do not know the role of cardioselectivity, intrinsic sympathonametic activity and membrane stabilizing activity in terms of efficacy in the treatment of superventricular arrhythmias. It would be useful to determine the potential advantages and disadvantages of each of these properties of the beta-blockers. Class III agents: the

FIGURE 4.

POTENTIAL FOR PHARMACOLOGIC THERAPY

	NEED
CLASS I A & C	REDUCED TOXICITY ON CHRONIC ADMINISTRATION
CLASS II	(A) CARDIOSELECTIVE EFFECT (B) ISA & EFFICACY (C) ROLE OF MEMBRANE ACTIVITY
CLASS III	(A) AGENTS OF REDUCED TOXICITY ON CHRONIC ADMINISTRATION (B) EFFICACY OF D-SOTALOL, NAPA, ETC.
CLASS IV	(A) ULTRA SHORT ACTING (B) DISASSOCIATE NEGATIVE DORMOTROPIC FROM NEGATIVE INOTROPIC ACTION.

amniodarones, sotolol, NAPA type agents seem of great promise in the treatment of superventricular arrhythmias. This potential should be thoroughly evaluated. There is also a distinct need for less toxic amniodarone-like drugs that are very useful in the treatment of superventricular arrhythmias, but due to their toxicity are severely limited. The acute intravenous ability of Class IV calcium blocking agents have revolutionized the acute treatment of superventricular arrhythmias. Still there is potential for ultra-short acting calcium blcokers in the treatment of superventricular arrhythmias as well as long acting, once a day preparations for chronic therapy. It would be useful to dissassociate the pharmacodynamic negative dormotropic effects from the negative inotropic action of the calcium channel blockers. Currently, one of the great limitations in the treatment of superventricular arrhythmias is that they are often associated with depressed myocardial function excluding the use of calcium blockers and beta blockers. Thus, therapy is left to the approach available the past two hundred years, the digitalis glycosides with the associated toxicities they offer.

SUMMARY

The evaluation of antiarrhythmic agents for the most part is focused on the treatment of ventricular arrhythmias. The advent of new techniques, especially invasive electrophysiologic testing combined with many new and exciting chemical entities has lead to an explosive effort to control ventricular arrhythmias and hopefully reducing sudden cardiac death. Development of agents for superventricular arrhythmias has, by necessity, played a secondary role. However, given the enormity of the problem and for the first time the availability of agents with potentially tolerable side-effects, the testing of drugs for the treatment of superventricular arrhythmias is becoming more frequent. Hopefully, this testing process will be systematic, employ both noninvasive and invasive techniques and carefully stratify the patient populations for arrhythmia etiology, type and the degree of underlying cardiac enlargement. The development of an effective antiarrhythmic arnementarium that patients can tolerate for the treatment of superventricular arrhythmias will be an important contribution to cardiovascular therapy.

PANEL DISCUSSION ON EVALUATION OF ATRIAL ARRHYTHMIAS

Dr. Morganroth: When one looks at an NDA for ventricular arrhythmias for efficacy, a few hundred patients are usually in the data base, does a ball-park figure exist for atrial arrhythmias?

Dr. Temple: The patients in an NDA serve several different functions. Showing effectiveness for intravenous verapamil in superventricular tachycardias was accomplished with 150 to 200 patients. There were many with atrial fibrillation so that may be more patients than you really need. That aspect of it wasn't very hard and I guess we had some reassurance from a long history of European use that provided some evidence of how well it was going to be tolerated. Looking at a drug with no other experience, you probably would not be comfortable with only 100 or 200 patients, because you wouldn't have much capacity to detect safety problems.

Dr. Morganroth: If you are evaluating a drug that has predominantly been studied for ventricular arrhythmias and you want to look at atrial arrhythmias and you were satisfied that the main data base answered questions of safety, so safety was not an issue, per se, might not 30 or 40 patients be sufficient for an efficacy claim? I say this since some of the data you showed on I.V. verapamil had only 30 or so patients.

Dr. Temple: I think that effectiveness can be established if the difference is clear in populations of that kind and that is not that unusual. The number of patients in control trials for anti-angina drugs that we have approved in recent years is well under 200. It is hard to do angina studies so you don't find many patients. What you do get is a lot more people treated in open trials, so I think the effectiveness aspect of it could probably be shown if the effect is reasonably large in a modest number of patients in each arrhythmia. You have got to break them apart. The safety concern is harder to deal with and I don't have a quick answer to that. I think what you are interested in is more open observational kinds of studies, and I wonder whether EP testing can't play a role in making that easier. I don't have a quick answer.

Dr. Moore: Dr. Watanabe, you made a comment that superventricular rhythms seem to be worsened by sympathetic stimulation but actually it has been my impression that with vagal stimulation, that the atrial arrhythmia tends to speed up. For example, atrial flutter can be converted to atrial fibrillation. As far as the ventricular response with vagal stimulation, goes it indeed is slowed. With vagal stimulation you tend to get a more rapid atrial rhythm.

Dr. Watanabe: I would agree with you in the case of atrial flutter, I think with vagal stimulation, you would convert that to atrial fibrillation. Therefore by definition, your atrial rate would be faster. You would go from 300 to 600 or whatever. But because of concealed conduction, actually the ventricular response may actually be slower and then with continued vagal tone on the AV node one would tend to get even a slower ventricular response. As I said, I tried to be careful in saying in most cases, there is a syndrome which I am sure some of the clinical electrophysiologists are much better aware of than I am of patients who have paroxysmal atrial fibrillation that seems to be vagally induced and as I understand, one of the characteristics of it is that it usually occurs at night. If these patients are Holter monitored often, their sinus rhythm actually slows down for a period of time before they develop paroxsymal atrial fibrillation. It can be shown that if one blocks the muscarinic receptor you can prevent this from happening. So that is one example in which vagal stimulation may actually cause the superventricular tachyarrhythmia. I think in most cases turning the vagus on, will either cardiovert the rhythm in the case of PSVT for example that uses the AV node as one of the limbs of the reentrant loop or can slow the ventricular response because of the effect on the AV node.

Dr. Katz: Clearly in the animal laboratory, the fibrillatory threshold of the atrium drops way down when you stimulate the vagus and it is much easier to prevent that fibrillation if you have a vagolytic state, and so it very well may be that disopyramide, given the fact it has a negative effect on the AV node is the best antifibrillatory drug for atrial fibrillation or quinidine secondarily as opposed to procainamide.

Dr. Watanabe: Do you mean because of its effect on the muscarinic receptor counteracting the vagal effect on the atrium to reduce fibrillatory threshold?

Dr. Katz: Yes.

Dr. Watanabe: The problem is that those drugs also will counteract the effect of the vagus on AV conduction and therefore as you know unless the patient is digitalized, the patient with atrial fibrillation will actually speed the ventricular response.

Dr. Somberg: Can I just comment on that in terms of the administration acutely of intravenously quinidine versus procainamide. Procainamide is far more likely to slow or abort the atrial fibrillation than is intravenous quinidine. We have not tried that with intravenous disopyramide. It is not available to us for testing, but I can tell you that we have just noted that in a number of patients we have been administering the drug to who were in fibrillation who were also undergoing stimulation studies and that quinidine can be a classic problem in that instance. I think whatever the hypothesis, you have to go back and figure out the theory to fit the clinical model. Both quinidine and disopyramide can accelerate the rate while procainamide might be more beneficial.

Dr. Ward: I would like to ask Dr. Temple to expand a bit on the issue of stratification of patients. Specifically, I am discussing those most difficult types of studies. Studies which show that long term prophylaxsis decreases the frequency or duration of attacks. Should the sponsoring company do two well controlled multicenter trials in WPW, AV nodal reentry tachycardia, MAT, EAT, etc. If so, that is a very large task. We saw some results from the verapamil long term study (Dr. Prichett), who did not define all these patients except that the EKG showed a narrow QRS regular tachycardia not based on an electrophysiologic diagnosis. I am wondering if there can't be a certain amount of lumping in at least some of your studies, with specific electrophysiologic studies to determine the mechanisms of action in some selected patients.

Dr. Temple: Anytime you include multiple forms of a particular kind of disease, you run the risk that there might be an effect in one of them that is different from the others. Just to make a time-honored example, we took an edema population and studied digitalis in it. If you had a lot of people whose edema was renal you would have made a mistake, you might not see an effect. If all of the superventricular arrhythmias responded more or less similarly, you would probably learn at least that in an unselected group of paroxsymal superventricular tachycardias, the drug worked, that might be an interesting finding. Most people would probably then ask whether there is some difference. It depends on the mechanism. I guess my reaction would be if you know a disease has several different subsets, it is wisest to study the subset if you can. It is safer for the company because you are less likely to make a mistake because you put the wrong patients in and you will know why a study either does or doesn't work out. Whether you could instead of doing that, do some sort of electrophysiologic screening that might suggest effectiveness in a particular condition and then test that by doing a chronic controlled trial, I am not sure. That seems an alternative to doing multiple studies in subsets.

Dr. Ward: I may not have explained my question well enough. I am not suggesting analyzing the patients together. What I am asking is almost a regulatory issue, does one need two well controlled studies in each of the arrhythmia subsets to gain an indication as opposed to a trend in one and a statistically significant results in another type in the same trial? Is that enough to get it approved for one or for superventricular arrhythmias including those in which the trends were found. That is what I am after.

Dr. Temple: That is a fair question. I guess my inclination is to say that if you have a set of closely related conditions, you may not need to do a pair of studies in each one of them and that the evidence ought to be considered additive in cases like that. I am not sure you need one in each if they are closely related. Let me add one thing. The one pure historical example we can site is that in the case of intravenous verapamil to convert superventricular arrhythmias, we accepted multi-center studies that lumped patients who came in with PAT, whatever the mechanism and noted that the conversion rates were approximately similar in both subgroups and we were satisfied with that. Maybe that was naive of us, but we were.

Dr. Horowitz: It is not such a big clinical problem because the incidence of the different mechanisms is so overwhelmingly tipped toward AV nodal reentry and accessory pathway SVT's that you don't have to worry about the rather rare example of sinus node reentry or intraatrial reentry or automatic atrial tachycardia in that population. Those types of patients are either very rare or rarely walk into the emergency room and say I am having frequent tachycardias, so that although there are 5 or 6 different mechanisms that are known, the distribution of those different mechanisms in a patient population is really skued and yet a pretty uniform group of patients.

Dr. Temple: The only distinction we have tried to make was between WPW and the others.

Dr. Winkle: I wanted to make a comment about Dr. Temple's presentation. Although I agree with almost everything he said, the one thing he said that did disturb me a little bit is the proposal for doing dose ranging by randomizing patients initially to a high dose of a drug, because I think none of us in clinical practice would want to overtreat the patient. We might randomize the patient to the high dose of verapamil and end up with a ventricular response of 30. The other question I wanted to ask is whether looking at the patient's symptomatic response to VPC suppression as a possible end point is adequate for showing drug efficacy, because there is no doubt that there is a sub-population of 20% or 25% of patients with frequent ectopic beats who really are quite symptomatic and a smaller percentage who are even almost disabled with these extra beats. They feel much better when the VPC's are almost eliminated (such as flecainide or encainide) and I wondered what the FDA's view would be if somebody came in and asked for an indication for a drug for making a patient feel better?

Dr. Temple: My presumption always is that if you are concerned about the possible side effects of the high dose, you would carry out that kind of study by using the low dose first followed by the high dose to make sure the low dose was tolerated. The point would be that you would have defined ahead of time the final dose, but you can reach it in stages, that is perfectly O.K. and if there is any concern of the kind you raised, that would be the way to do it. The second thing is, if someone were to seek a control of for an APc indication, I would think that one of the pieces of evidence that would be brought to bear, that that is a reasonable thing to do , or that you don't need all that much additional evidence of safety or something would be that people would be accessed for their symptomatic improvement. We probably would like to see more explicit attention to quality of life measurements that everyone is talking about and I would think that is a legitimate reason for treatment.

Dr. Winkle: You said APc, what about VPC?

Dr. Temple: Same goes.

Dr. Pratt: I have one comment, two questions. Dr. Temple was talking about the risk of atrial fibrillation post MI using the Framingham data. I think that atrial fibrillation in that study and most of the epidemiologic studies probably is a marker for the people who are on digitalis and I think we can't forget that digitalis is really one of the major reasons that there was an increased mortality. I would like any comments would have. Number two. All of you have Holter monitoring data on your slides. None of you told us how to use them. Nobody liked the idea of counting PAC's, you all thought that was irrelevant, and if it is irrelevant relative to other more serious atrial arrhythmias such as fibrillation and flutter, how in the world do you use Holter monitoring in any of those studies.

Dr. Somberg: The implication is you have to factor that into the specific arrhythmia you are studying and what your goals are. I mean there are obvious things if you are looking at somebody who is in atrial fibrillation who gets cardioverted and then goes on quinidine or experimental drug for instance there, you do the Holter monitors to find out if they are all the time in sinus rhythm. Periodically you might do it at 1 week, 8 weeks, 12 weeks, 18 and so on and it is a way to look at more than a 12 lead EKG which samples only 220 heart beats. That is one way to do it. The other way is to use 24 hour Holters at a certain point repeatedly to see how you are doing versus placebo or something else to catch a slice a life, so to speak, and compare that to the intervention studied or to use event recorders, which would trigger if they are worn for 3 to 5 days and to have the patient activate it at certain periods of time.

Dr. Pratt: I would like to pin you down a little more though, because in the first case in paroxsymal atrial fibrillation, you don't need to do Holter monitors, you could just do an event recorder everyday and do the same thing. I don't know how you quantitate it. You haven't told me anything about quantitation and it is a very qualitative - yes-no-here for 3 hours out of the last 48 hours. Is there any way that that is approachable or is it really not very useful?

Dr. Somberg: I think there are ways it is approachable and they are complementary (Event recorders and Holter monitoring). Manytimes people have a slight change in their arrhythmi or have side effects due to a medication and become less perceptive of their arrhythmia. So just going by what they press the button for, and there is also, I mean behavioral pscyhologists will tell you, you get very interested, you press the button for a lot of things, and then all of a sudden in a 24 hour to 5 day period you stop pressing the event button unless it is either a bothersome event or something that is hemodynamically unstable. So I think knowing what happens for a full 24 hours is essential and just to throw out the Holter monitor is wrong. It depends also on what arrhythmia you are looking at? For atrial fibrillation I would say I would like to know the number of hours in sinus rhythm, the number of hours or minutes in an atrial arrhythmia, the rate of that atrial arrhythmia, and the number of times one is in it. I think these are all important factors if going in and out of atrial fibrillation on occasions increases the propensity towards embolization, the patient may be better off left in atrial fibrillation with a controlled response most of the time. I think those would be important things to quantify.

Dr. Temple: The studies that are underway with verapamil in prevention of paroxysmal superventricular tachycardias chose in some cases to use people who had a lot of events so that in any 24 hour period, you would be sure to find lots of them, and in that case you can analyze the Holter in a sort of routine way. Set any standard you want. More than 3 beats, more than 10 beats, and that is anybody's choice. Where they are relatively rare, Holter monitoring is probably not going to be much help except for the safety analysis. If you induced anything bad, and I think there you just have to go to a patient defined event, whether it is a diary or pushing a button to record what it is or getting to the telephone or something like that.

Dr. Nacarelli: I just wanted to add that I think most of what we have gone over with the event recorders, the EP studies, I would be in agreement with. I think that with whatever study is going to be designed, Holter monitoring is going to be a necessary part of it. Now it depends on which patient group you are talking about. Obviously most patients with PSVT have very rare atrial ectopy; they may have one PAC a day for three months and then have an episode of SVT but I think it is worth getting 48 hours of control Holtering to identify that population before-hand. There are other patients that have frequent atrial ectopy, especially groups with ectopic atrial tachycardia, 30,000 PAC's a day, 300 non-sustained runs of SVT, they are more bothered by their palpitations and symptoms and in those patients, maybe the purist end point may be Holter monitoring in conjunction with a cardiobeeper, patient diary which maybe your final evaluation end point. We don't know the variability of atrial arrhythmias. Studies haven't been done like Dr. Morganroth has done with ventricular arrhythmias, but I think that probably some kind of baseline atrial Holter monitoring is going to have to be done. That is going to help you substratify your patients and it may be used as the end point analysis and also identify proarrhythmias.

Dr. Morganroth: Dr. Temple, it seems to me that if you look at digitalis, all the data suggests that it is potentially very deleterious in certain commonly used groups (eg. patients post-infarction, those with atrial fibrillation with other concomitant drugs as quinidine, etc.) The data on efficacy of digitalis except for rate control in atrial fibrillation (for which we now have many alternative drugs) is poor so why has the FDA still continued to have digitalis on the market?

Dr. Temple: There is in fact some evidence that it is effective as an inotrope. Certainly not effective in prolonging life. There isn't any evidence that any treatments with heart failure do that in the larger sense. But there are studies that are small but that do show that it is an effective inotrope and that it prevents heart failure. It is very hard to access things like whether being on a drug increases or decreases mortality and the kind of data we have are of uncertain ability to show that. It certainly is possible, I guess I would need to read it again. I am not sure that Kannel's data don't correct for atrial fibrillation case. I'm not sure they don't correct for digitalis use. I would have thought they would have, but I am not sure the paper is explicit on that question. Certainly a sophisticated epidemiologist would know that that it is something he ought to correct for. It is very hard to know whether being on digitalis or not on digitalis is good for you or bad for you. That is not an easy thing to find out without a control trial of enormous proportions.

Dr. Pratt: The reason that it is quite a relevant question right now as most of us in the room right know, is the NIH has put out an RFP for heart failure. There is going to be a National trial looking at mortality and heart failure and everybody is going to get on one side or the other, there is going to be half the people in the room with a religious feeling for digitalis. And we know they are going to pick patients with class 3 and 4 heart failure and those are the people most likely to get digitoxic. I think it is going to be a very interesting discussion and it is too bad we don't already have the data that you surmise that we need to answer the question.

Dr. Somberg: There is something important to be said about the use of digitalis for superventricular arrhythmias, the topic of the afternoon, is that unfortunately, we do have a lot of agents to treat acute problems, but at least I would say 30-50% of patients who present with hemodynamically compromising rapid atrial fibrillation, because of either very poor LV function or because we see less and less rheumatics and more heart failure patients because of very poor LV function of because of compromising COPD or maybe asthma, the treatments of beta blockade and verapamil are much less turned to and one tends to turn to digitalis. For the control of ventricular response, I think digitalis still has a place and we would be very hard pressed if digitalis was taken away from us to treat these sort of patients.

Dr. Temple: Just a personal opinion based on available data. It would certainly not be my view that everybody in heart failure study has to have whatever is being tested added to digoxin. There really definitely ought to be a no digitalis group. I don't know if there ought to be a no diuretic group, but I know what my answer to that study question would be.

Dr. Pratt: I would like to refer to what you just talked about. We are talking about atrial arrhythmias and yesterday we presented 31 cases of torsade and 3/4 of the cases of torsade came from quinidine treatment of atrial tachyarrhythmias, not ventricular arrhythmias, so I think it is a very relevant question to this discussion.

SYMPOSIUM ON NEW DRUGS AND DEVICES

October 4 & 5, 1984

Participant List

William B. Abrams, M.D.
Executive Director, Scientific Development
Merck Sharp & Dohme Research Laboratories

Mahmood Alikhan, M.D.
Director, Cardiac Floor and Cardiac Rehabilitation Program
Greater Baltimore Medical Center

Olle Almgren, M.D., Ph.D.
Cardiovascular Pharmacology
AB Hassle, Sweden

Cynthia B. Altman, M.D.
Director, Worldwide Clinical Research Management
Smith Kline & French Laboratories

Kelley Anderson, M.D.
Division of Cardiology
University of Utah

Michael J. Antonaccio, Ph.D.
Vice President, Cardiovascular Research and Development
Bristol-Myers Company

Mirza M. A. Beg, M.D.
Vice President, Planning & Operations, Medical Affairs
Smith Kline & French Laboratories

Prof. Dr. Franz Bender
Department of Cardiology
West Germany

J. Thomas Bigger, Jr., M.D.
Professor of Medicine and of Pharmacology
Columbia University

Bernard H. Boal, M.D.
Clinical Associate Professor of Medicine
New York Medical College

Michael J. Borland, M.D., Ph.D.
Associate Medical Director
E. I. du Pont de Nemours & Co., Inc.

Agnes A. Bouvet, M.D.
Associate Director, Cardiovascular Clinical Research
Ciba-Geigy Corporation

George B. Bren, M.D.
Assistant Professor of Medicine
The George Washington University Medical Center

James A. Brinckerhoff
Office of Medical Research
A. H. Robins Company, Inc.

Robert S. Brown
Director, New Studies
Cardio Data Systems

Larry R. Bush, Ph.D.
Senior Research Pharmacologist
Merck Institute for Therapeutic Research

William J. Cady, Pharm.D.
Manager, Clinical Research
Marion Laboratories, Inc.

Ronald W. F. Campbell, M.D.
Senior Lecturer in Cardiology
Freeman Hospital, England

Rita A. Carey, Ph.D.
Scientific Administrator, Medical Affairs
Smith Kline & French Laboratories

Joan Carter
Vice President
United Medical Corporation

Peter Cervoni, Ph.D.
Head, Cardiovascular Biological Research
American Cyanamid Company

Peter Chan, Ph.D.
Senior Research Pharmacologist
American Cyanamid Company

Sughok K. Chun, M.D.
Food & Drug Administration
Maryland

Jaime B. Coelho, M.D.
Director, Cardiovascular Clinical Research
Ayerst Laboratories

Jay N. Cohn, M.D.
Professor of Medicine
University of Minnesota

Glenn G. Cousins
Research Division Manager
Cardio Data Systems

Bruce P. Damiano, Ph.D.
Research Scientist, Cardiovascular Research
McNeil Pharmaceutical

Michael Davidov, M.D.
Clinical Assistant Professor
Georgetown University School of Medicine

Jeff Davies
Manager, New Product Planning - MIS Division
Eli Lilly and Company

Richard R. Dean, Ph.D.
Assistant Director, CV/Renal Clinical Research
G. D. Searle & Co.

Robert DiBianco, M.D.
Cardiology Research
Veterans Administration Medical Center

Federico Dies, M.D., Ph.D.
Associate Clinical Investigator
Eli Lilly and Company

Martin Duke, M.D.
Associate Director, Clinical Investigation
Boehringer Ingelheim Ltd.

Edel Berman, M.D.
Bristol-Myers Co.
New York

Stewart J. Ehrreich, Ph.D.
Deputy Director, Division of Cardio-Renal Drugs
Food and Drug Administration

Carol Ellis, M.D.
McNeil Pharmaceutical

Toby R. Engel, M.D.
Professor of Medicine
Medical College of Pennsylvania

Daneil M. Estes, R.N.
Clinical Research Assistant
University of Miami

David A. Evans
Systems Analyst
Research Data Corporation

Alfred F. Fasola, Ph.D., M.D.
Senior Clinical Pharmacologist
Lilly Laboratory for Clinical Research

John L. Fischetti, M.D.
Associate Director of Medical Affairs
Stuart Pharmaceuticals

John D. Fisher, M.D.
Associate Professor of Medicine
Albert Einstein College of Medicine

Richard N. Fogoros, M.D.
Assistant Professor of Medicine
University of Pittsburgh School of Medicine

Michael J. Fox, M.D.
Associate Director, Medical Affairs
Smith Kline & French Laboratories

Richard P. Friedlander, M.D.
Director of Cardiovascular Research
Janssen Pharmaceutica, Inc.

John A. Gillespie, M.D.
Director, Cardiology Unit/Medical Intensive Care Unit
Highland Hospital, New York

Frederick Ginsberg, M.D.
Brachfeld Medical Associates
New Jersey

Kenneth M. Given, M.D.
Executive Director, Regulatory Affairs
Merck Sharp & Dohme Research Laboratories

Fredarick L. Gobel, M.D.
Associate Professor of Medicine
University of Minnesota Medical School

Michael B. Goodkin, M.D.
Upland, Pennsylvania

Richard Gorlin, M.D.
Murray M. Rosenberg Professor
The Mount Sinai Medical Center

Thomas B. Grayboys, M.D.
Assistant Professor of Medicine
Brigham & Women's Hospital, Massachusetts

Richard M. Greenberg, M.D.
Clinical Assistant Professor of Medicine
Thomas Jefferson University Hospital

H. Leon Greene, M.D.
Associate Professor of Medicine
University of Washington

Allan M. Greenspan, M.D.
Associate Professor of Medicine
Hahnemann University

Arnold J. Greenspon, M.D.
Associate Professor of Medicine
Thomas Jefferson University Hospital

Lawrence S. C. Griffith, M.D.
Associate Professor of Medicine
The Johns Hopkins Hospital

H. F. Hailman, M.D., Ph.D.
Director of Scientific Liaison
Hoffmann-LaRoche Inc.

Dr. Bryan J. Harlow
Medical Advisor
Syntex Research, England

Donald C. Harrison, M.D.
William G. Irwin Professor of Cardiology
Stanford University School of Medicine

Richard H. Helfant, M.D.
Professor of Clinical Medicine
University of Pennsylvania School of Medicine

Marc M.J. Henis, M.D.
Executive Director, Cardiovascular Clinical Research
Ciba-Geigy Corporation

John Hermanovich, M.D.
Assistant Professor of Medicine
Hershey Medical Center

Carol Hinzman, R.N.
Washington D. C. Veterans Administration Medical College
Washington, D. C.

Brian F. Hoffman, M.D.
David Hosack Professor
College of Physicians & Surgeons, Columbia University

Geoffrey I. Holmes
Assistant Director, Clinical Research - CV
Merck Sharp & Dohme Research Laboratories

Leonard N. Horowitz, M.D.
Associate Professor of Medicine
Hahnemann University

Ping-Hwa Hsu, Ph.D.
Senior Research Scientist
Bristol-Myers Company

Edward M. Hughes, M.D.
Associate Director, Clinical Research
Boehringer Ingelheim Ltd.

Margaret E. Hurley, M.D.
Director, Cardiovascular Clinical Research
Ciba-Geigy Corporation

John D. Irvin, M.D., Ph.D.
Senior Director, Clinical Research - CV
Merck Sharp & Dohme Research Laboratories

Dorte Jensen
Medical Writer
Janssen Pharmaceutica

M. King Jolly, Pharm. D.
Clinical Trials Specialist
Burroughs Wellcome Co.

Dietrich Kalusche, M.D.
Rehabilitationszentrum
West Germany

Ronald Kartzinel, M.D., Ph.D.
Deputy Director, Medical Research
E. I. du Pont de Nemours & Co., Inc.

Richard J. Katz, M.D.
Associate Professor of Medicine
George Washington University School of Medicine

F. Bryan Kennedy, M.D.
Pittsburgh, Pennsylvania

Julie A. Kersten
Research Assistant
The Upjohn Company

Richard C. Klein, M.D.
Assistant Professor of Medicine
University of New Mexico

S. A. Kolenik
Vice President
Intec Systems, Inc.

Gregory A. Kopia, Ph.D.
Associate Senior Investigator, Cardio-Renal Pharm.
Smith Kline & French Laboratories

John L. Kovacs
Senior Clinical Research Associate
Hoffmann-LaRoche Inc.

Ezra Lamdin, M.D.
Director, Medical Affairs

Stuart Pharmaceuticals
Diane Lawthorne, R.N.
Washington D. C. Veterans Administration Medical College
Washington, D. C.

Ralph Lazzara, M.D.
Professor of Medicine
University of Oklahoma Health Sciences Center

William Likoff, M.D.
Professor of Medicine, Chancellor
Hahnemann University

Sheng K. Lin, Ph.D., M.D.
Assistant Director of Clinical Pharmacology
Revlon Health Care Group

Donna Lindemuth, R.N.
Washington D. C. Veterans Administration Medical College
Washington, D. C.

Raymond J. Lipicky, M.D.
Acting Director, Division of Cardio-Renal Drug Products
Food and Drug Administration

David T. Lowenthal, M.D.
Professor of Medicine and Pharmacology
Likoff Cardiovascular Institute of Hahnemann University

Robert R. Luther, M.D.
Director of Clinical Research, Cardiovascular Section
Abbott Laboratories

Sara Armstrong Mahler, M.D.
Group Director, Cardiovascular Diseases
E. I. du Pont de Nemours & Company, Inc.

Kim Mays-Whitlen, Pharm. D.
ICI Americas, Inc.

Gay McAllister
Techical Director, Ambulatory Monitoring
Survival Technology, Inc.

Valerie J. McCall
Operations Manager
Cardio Data Systems

Bernard F. McDonagh, Ph.D.
Market Development Manager
Riker Laboratories, Inc./3M

Charles F. McNally, M.D.
Director, Cardiovascular/Renal Group
Smith Kline & French Laboratories

Edwin F. McNichols, M.D.
Harleysville, Pennsylvania

Howard Miller, M.D.
Medical Director
Sandoz Research Institute

Morton Mower
The Johns Hopkins University School of Medicine

E. Neil Moore, D.V.M., Ph.D.
Professor of Physiology in Medicine
University of Pennsylvania

Joel Morganroth, M.D.
Professor of Medicine and Pharmacology
Hahnemann University

Susan Moser, R.N.
Clinical Research Representative
Intec Systems, Inc.

Gerald V. Naccarelli, M.D.
Assistant Professor of Medicine
University of Texas Medical School at Houston

Michael O'Reilly, M.D.
ICI Americas, Inc.

Charlotte Anne Panis
Clinical Research Scientist
Sandoz Resarch Institute

Indravadan I. Patel, M.D.
Research Physician
Hoffmann-LaRoche Inc.

Gerd Petrik
Helopharm KG
Germany

William S. Pickens, M.D.
Pensacola, Florida

Gordon W. Pledger, Ph.D.
Group Leader, Statistical Evaluation Branch
Division of Biometrics, FDA

Craig M. Pratt, M.D.
Assistant Professor of Medicine
Baylor College of Medicine

Eric N. Prystowsky, M.D.
Associate Professor of Medicine
Indiana University School of Medicine

Glenn A. Rahmoeller
Acting Director, Division of Cardiovascular Devices
Food and Drug Administration

Robert Reder, M.D.
Director, Medical Affairs
Knoll Pharmaceuticals

Phillip Reid, M.D.
The Johns Hopkins Medical Center
Baltimore, Maryland

Joseph Reiser, M.D.
Berlex Laboratories

Robert D. Reynolds, Ph.D.
Senior Research Fellow
American Critical Care

R. Linda Roemer
Director, Science Information
Janssen Pharmaceutica

Sandra B. Rohan, R.N.
Pensacola, Florida

Alberto Rosenberg, M.D.
Vice President, Clinical Research
Stuart Pharmaceuticals, Inc.

Magdi Sami, M.D.
Assistant Professor of Medicine
McGill University

Stephen B. Schectman
President
Research Data Corporation

C. Richard Schott, M.D.
Upland, Pennsylvania

Robert Sheehan
Boehringer Ingelheim

Steven Singh, M.D.
Cardiology Department
Washington D.C. Veterans Adm. Medical College

Maria Smith
Clinical Research Associate
Hoffmann-LaRoche Inc.

Peter Somani, M.D.
Professor of Medicine & Pharmacology
Medical College of Ohio

John Somberg, M.D.
Assistant Professor of Medicine
Albert Einstein College of Medicine

Lester Soyka, M.D.
Bristol-Myers Company

Scott R. Spielman, M.D.
Associate Professor of Medicine
Hahnemann University

Lois A. Steele
ICI Americas, Inc.

Mitchell Steinberg, Ph.D.
Cardiovascular Pharmacology Research Administration
Eli Lilly and Company

Janice L. Stickney, Ph.D.
Associate Director, Office of Scientific Affairs
G. D. Searle & Company

David M. Stout, Ph.D.
Group Leader, Medicinal/Organic Chemistry
American Critical Care

Mark Sullivan, Ph.D.
Berlex Pharmaceuticals

Max W. Talbott, Ph.D.
Regulatory Affairs Associate
Eli Lilly and Company

Robert Temple, M.D.
Acting Director, Office of Drug Research & Review
Food and Drug Administration

Udho Thadani, M.D.
Professor of Medicine
University of Oklahoma Health Sciences Center

Ronald J. Trancik, Ph.D.
Clinical Project Manager
Riker Laboratories

Eugenia Triantas, M.D.
Food and Drug Administration

Ilhan H. Tuzel, M.D.
Associate Director Cardiovascular/Renal Research
Hoffmann-LaRoche Inc.

Michael J. Umen, Ph.D.
Philadelphia, Pennsylvania

P. Jacob Varghese, M.D.
Professor of Medicine and Pediatrics
The George Washington University Medical Center

P. Verdow, M.D.
Warner-Lambert International
The Netherlands

David A. Vondle
Manager, Corporate Pharmaceutical New Product Planning
Eli Lilly and Company

Richard Wade
Director, Cardiovascular Portfolio Management
Ciba-Geigy Corporation

David P. Ward, M.D.
Assistant Director, Clinical Research
Riker Laboratories, Inc./3M

August M. Watanabe, M.D.
Professor of Pharmacology
Indiana University Medical Center

Richard E. Weber, M.D.
Director, Cardiovascular Research
Warner Lambert/Parke-Davis

Anne Wierman
Houston, Texas

Park W. Willis, III, M.D.
Professor of Medicine
Michigan State University

Roger A. Winkle, M.D.
Associate Professor of Medicine
Stanford University Medical Center

Florence Wong, Pharm. D.
Riker/3M

Samuel Wong, Ph.D.
Group Leader, In Vitro Antiarrhythmic and Cardiotonic
Berlex Laboratories

Raymond L. Woosley, M.D., Ph.D.
Associate Professor of Medicine and Pharmacology
Vanderbilt University Medical Center

Milford G. Wyman, M.D.
Clinical Professor of Medicine
University of Southern California School of Medicine

Patricia A. Zempol
Senior Clinical Research Associate
Hoffmann-LaRoche